A MANIFESTATION OF MONSTERS

EXAMINING THE (UN)USUAL SUSPECTS

DR. KARL P.N. SHUKER

ANOMALIST BOOKS
*San Antonio * Charlottesville*

A Manifestation Of Monsters: Examining The (Un)Usual Suspects

Cover art: Cryptids © Michael J. Smith

Book design by Seale Studios

For information, go to AnomalistBooks.com, or write to:
Anomalist Books, 5150 Broadway #108, San Antonio, TX 78209

Some of the chapters or sections of chapters in this book are reprinted, expanded, updated, or amalgamated versions of the following previously published articles or *ShukerNature* blog posts:

"On the Track of the Nandi Bear," *Fortean Times*, No. 315 (June 2014): 46-49.

"Bilbo Baggins Versus the Mongolian Death Worm?" *ShukerNature*, http://www.karlshuker. blogspot.co.uk/2013/11/bilbo-baggins-versus-mongolian-death.html 6 November 2013.

"Monitoring Mystery Varanids, Part 1," *Practical Reptile Keeping*, May 2013: 56-60.

"Monitoring Mystery Varanids, Part 2," *Practical Reptile Keeping*, June 2013: 56-60.

"Monitoring Mystery Varanids, Part 3," *Practical Reptile Keeping*, July 2013: 56-60.

"Gigantic Geckos," *Practical Reptile Keeping*, December 2014: 56-60.

"The Camp Fircom Caddy Carcase—Monster or Montage?—Reviewing a Little-Known Sea Serpent Controversy," *ShukerNature*, http://www.karlshuker.blogspot.co.uk/2015/03/the-camp-fircom-caddy-carcase-monster.html 9 March 2015.

"The Curious Case of Rothschild's Lost Tusk and the Non-Existent Elephant Pig—An Enduring Cryptozoological Conundrum From Africa," *ShukerNature*, http://www. karlshuker.blogspot.co.uk/2014/07/the-curious-case-of-rothschilds-lost.html 3 July 2014.

"Gigantic Frogs and Toads, Part 1," *Practical Reptile Keeping*, May 2015: 56-60.

"Gigantic Frogs and Toads, Part 2," *Practical Reptile Keeping*, June 2015: 56-61.

"The Striped Anteater That Made a Buffoon Out of Buffon," *ShukerNature*, http://www. karlshuker.blogspot.co.uk/2014/03/the-striped-anteater-that-made-buffoon.html 13 March 2014.

"A Baby Chupacabra?—You're Having a Laugh—Or a Gaff!!" *ShukerNature*, http://www. karlshuker.blogspot.co.uk/2013/01/a-baby-chupacabra-youre-having-laugh-or.html 23 January 2013.

"Spiders With Wings," *Practical Reptile Keeping*, February 2015: 56-59.

"Lairs of the Lizard-Men," *Paranormal*, No. 50 (August 2010): 14-19.

"Exposing the 'Dead Bigfoot Photo'—The Bear Facts at Last!—A *ShukerNature* World Exclusive!" *ShukerNature*, http://www.karlshuker.blogspot.co.uk/2015/04/exposing-dead-Bigfoot-photograph-bear.html 11 April 2015.

"A Tale of Tangled Tails," *Small Furry Pets*, No. 18 (December 2014-January 2015): 44-49.

"The Vegetable Lamb of Tartary," *Small Furry Pets*, No. 19 (February-March 2015): 44-49.

"Oarfish Origins and a Very (Un?)Likely Sea Serpent," *ShukerNature*, http://www.karlshuker.blogspot.co.uk/2014/12/oarfish-origins-and-very-unlikely-sea.html 26 December 2014.

"Was a Golden Freshwater Oarfish Encountered During the Vietnam War?" *ShukerNature*, http://karlshuker.blogspot.co.uk/2015/06/was-golden-freshwater-oarfish.html 7 June 2015.

"Why Blood-Drained Carcases are Not the Work of Chupacabras or Other Supposedly Vampiric Cryptids," *ShukerNature*, http://www.karlshuker.blogspot.co.uk/2013/07/why-blood-drained-carcases-are-not-work.html 19 July 2013.

"When Horned Rodents Walked the Earth," *Small Furry Pets*, No. 17 (October-November 2014): 42-47.

"Never Bother a Beithir," *Wild About Animals*, 8 (July 1996): 9.

"Bothersome Beithirs and Other Freshwater Mystery Eels," *Popular Fish Keeping*, Summer 2014: 62-65.

"Pigging Out at Christmas," *Fortean Times*, No. 296 (January 2013): 55-58.

"Strange Cases From the Caribbean," *Small Furry Pets*, (June-July 2013): 52-55.

"Blue Pigs and Hyaena-Dogs," *ShukerNature*, http://www.karlshuker.blogspot.co.uk/2011/03/blue-pigs-and-hyaena-dogs.html 12 March 2011.

"The Nung-Guama and the Nyamatsane—A Couple of Obscure Crypto-Primates From Once Long Ago?' *ShukerNature*, http://www.karlshuker.blogspot.co.uk/2011/04/nung-guama-and-nyamatsane-couple-of.html 8 April 2011.

"Mystery Insects of Many Kinds," *Practical Reptile Keeping*, April 2014: 56-60.

"The Beast of Buderim," *Wild About Animals*, 8 (May 1996): 9.

"Giant Blue Eels of the Ganges—Worming Out a Possible Explanation," *ShukerNature*, http://www.karlshuker.blogspot.co.uk/2014/12/giant-blue-eels-of-ganges-worming-out.html 27 December 2014.

CONTENTS

To my dear mother,
Mary Doreen Shuker (1921-2013)

If only my tears could be the footsteps
that lead me to where you now are.
If only my grief could light my way there,
a lantern with which I could see.
If only my faith could bear me upwards
to Heaven's bright portal afar.
If only my love could find you waiting,
and bring you back homeward with me.

For you, Mom, with all my love, always.

FOREWORD

When we were children, our parents used to assure us that there were no such things as monsters, unless of course they were determined to discourage us from wandering into the woods after dark. On those occasions, they would merely imply that there may be things in this world that are better off left alone, and end the conversation right there. Of course, some of us found out much later that this attitude was altogether fallacious, as our planet is literally teeming with monsters of all shapes and sizes. For millennia, our ancestors have been sitting around campfires and in hushed whispers describing those vile, grotesque creatures cursed to prowl the godforsaken regions of our world, and often capable of reducing even the most gallant explorer into a quivering coward.

Despite these traditions, we seem to be losing touch with our primordial wisdom. Modern psychologists have suggested that monsters are merely instruments of our imagination, a way for us to put a face on our deep-rooted fears by imparting them with form and substance. Tangible notions are, after all, much easier for us to wrap our minds around and confront. Yet the problem appears to be far more arcane, since there are indications that some of our so-called monsters are undeniably corporeal in nature. There are enduring legends, of course—too many to list, stemming from virtually every culture around the globe, and oftentimes astoundingly similar despite the great distances dividing them.

In addition, we have modern eyewitness accounts, and that's where things get interesting. You see, over many years I've gathered testimony from hundreds of seemingly credible persons who claim to have actually encountered living, breathing monsters. As a result of these terrifying experiences, the foundations of these peoples' lives have been irrevocably shaken. Just imagine the emotional impact of having a "myth" step right out in front of you! Admittedly, in the absence of any hard physical evidence for these remarkable allegations, it's tempting to dismiss the whole lot as a series of misidentifications, hallucinations, or outright whoppers, and in fact the skeptics have done just that. Notwithstanding, I invariably point to three important factors:

First off, our planet—while ostensibly shrinking on a daily basis as a result of astonishing advancements in technology and communications—is still largely unexplored. The majority of Earth's

Ken Gerhard in the Garibaldi Mountains of British Columbia, Canada (© Ken Gerhard)

surface is noticeably covered with very deep and inaccessible water, which even the most conservative scientists will acknowledge is capable of harboring sizeable, elusive animals. Moreover, at least half of the remaining land surface is composed of virgin wilderness—regions of impenetrable jungle and swamp, vast deserts, frozen tundra, and expansive mountain ranges that could hide a multitude of unknown beasts.

Accordingly, new life forms are still being discovered on a regular basis, and not all of them are microscopic. Some, in fact, have been truly substantial, and we shouldn't be too surprised to learn that many recently named species were considered fables until they unexpectedly wound up on a zoologist's examination table.

Finally, we have in many instances compelling clues as to the true nature of our monsters. Oftentimes the physical characteristics depicted in legends in addition to present-day descriptions align convincingly with those of genuine animals that existed in the past, or otherwise reflect a sensible ecological niche. So, the real question then becomes: Who can we trust to separate the wheat from the chaff?

For over a quarter of a century, Karl P.N. Shuker has been exceptionally skilful when it comes to injecting a degree of reason and order into the kingdom of enigmatic creatures. A rare level of expertise in pertinent disciplines including physiology, taxonomy, and global mythology arms him with a perspective that is impressive to say the least. Moreover, Karl radiates a youthful and unwavering enthusiasm for the subject matter that translates to pure joy for the reader. Through a cherished correspondence, I've personally benefited from his insight on more occasions than I can recall.

And this particular assemblage of monsters does not disappoint. To anyone acquainted with the field of cryptozoology, many of the names will be familiar—Caddy, Nandi bear, Mongolian death worm, and Jersey devil to name a few. But, as always, Karl is able to dig up obscure details revealing provocative clues completely overlooked by other investigators. In this edition, he manages to banish some longstanding misconceptions regarding several widely celebrated mysteries. Additionally, Karl displays an uncanny knack for bringing to light lesser-known cryptids that even seasoned researchers like myself have never heard of, and with this compendium he delivers once again. From Scotland's amphibious eel-like beithir, to Sweden's bat-winged nattravnen, to Iraqi marsh dragons, a blue-skinned pig from Nicaragua, and a myriad of others that will leave little doubt...monsters are real!

— Ken Gerhard

INTRODUCTION

Monsters come in all shapes and sizes. Some of them are things people are scared of. Some of them are things that look like things people used to be scared of a long time ago. Sometimes monsters are things people should be scared of, but they aren't.
— Neil Gaiman, *The Ocean at the End of the Lane*

There are things known and there are things unknown, and in between are the doors of perception.
— Unattributed modern-day quote inspired by Aldous Huxley's *The Doors of Perception* and by words spoken by Ray Manzarek of 1960s American rock band The Doors

Enshrouded by the mists within our minds are the doors of perception, of our perception, and no two doors are ever the same. Who can say what we shall encounter when we step through their shadowed archways? What lies waiting on the other side? There may be magic, there may well be mystery, and there may very well be monsters...

In the 30 years during which I have been investigating and documenting mystery creatures, my writings have been inspired by countless different influences, but what inspired this present book was a spectacular work of art—specifically, the wonderful illustration that graces its front cover, which was prepared by Michael J. Smith, an immensely talented artist from the USA who also happens to be a longstanding Facebook friend of mine.

It was back in early 2012 that Michael drew my attention to his painting entitled "Cryptids," which he had recently completed, and when I saw it I knew immediately that it would make a superb front-cover illustration for a book. I mentioned this to Michael, and he replied straight away that if I'd like to use it for that exact purpose with some future book of mine, I was more than welcome to do so.

Ever since then, even though I've been busy with many other writing projects, the notion of preparing a book inspired directly by Michael's painting and the 17 mysterious, controversial creatures depicted in it has always stayed with me, inciting all manner of ideas and schemes along the way concerning its possible scope, style, and, above all, its content.

Incidentally, just in case you are wondering what all of those creatures in Michael's painting are, as some are not particularly well known, here's a complete listing:

Top row, left to right: Loch Ness monster, giant squid, giant moa, mokele-mbembe.

Second row, left to right: bigfoot, mothman, Jersey devil.

Third row, left to right: skunk ape, dogman, pterodactyl, "Creature from the Black Lagoon"-type reptoid.

Fourth row, left to right: thylacine, chupacabra, dodo.

Front row, left to right: Mongolian death worm, Comoros coelacanth, Mexican mini-chupacabra.

The problem that I faced if I were to remain true to my ambition to include all of them within a single book was twofold.

Firstly: these particular creatures have been featured in countless previous books, and generally with the same familiar information about them endlessly recycled and regurgitated. So if they were also going to feature in mine, how could I document them in a different, original manner—was there unusual, obscure, but entertaining material out there for each one that had rarely if ever been assessed before? Although it took considerable time and research, I'm happy to say that, yes, there is such material out there, and, yes, I uncovered plenty of it.

Secondly: how could I categorize these creatures collectively, because viewed in total they represent an exceedingly diverse array of forms. Most of them are unquestionably cryptozoological, and thereby fall comfortably within the range that I normally deal with, but there are also certain ones that seemingly exist at that range's very furthest extent or even completely beyond it. These latter examples include the dogman, the two types of chupacabra, the Jersey devil, and the "Black Lagoon"-type reptoid. Even the pterodactyl has some ostensibly preternatural representatives (such as the Swedish nattravnen or night raven), the mothman is deemed to be paranormal or alien by some researchers, and so too for that matter are the skunk ape and even the bigfoot.

What single term could I use, therefore, that would effectively embrace, encompass, and enumerate all 17 members of this highly select, exclusive company, as well as the range of additional creatures that I would also be including in the book? "Cryptid" was not sufficiently comprehensive for my purpose, nor was "mystery creature" or "unknown animal." It finally dawned upon me that there was only one such term that could satisfy all of those requirements—indeed, it was tailor-made for such a role. The term? What else could it be? "Monster"!

Derived from the Latin noun "monstrum" and the Old French "monstre," down through the centuries "monster" has elicited a number of different definitions, although some of them overlap to a certain extent. Consequently, a monster can be a very strange, frightening, possibly evil (and/or ugly) mythical creature; something huge and/or threatening; a malformed, mutant, or abnormal animal specimen; and even something extraordinary, astonishing, incredible, unnatural, inexplicable. These definitions collectively cover all of this book's subjects—and so too, therefore, does the single word "monster" from which the definitions derive.

Thus it was that this became a book of monsters, but not just a book—a veritable manifestation of monsters. That is, a unique exhibition, a singular gathering, an exceptional congregation of some of the strangest, most mystifying, and sometimes truly terrifying creatures ever reported—still-unidentified, still-uncaptured, still-contentious. Even "mainstream" species like the dodo and coelacanth, whose reality and zoological identity are fully confirmed, still succeed in eliciting controversies, and are veritable monsters—the dodo having been referred to by various researchers as a monstrous dove, and the coelacanth as a resurrected prehistoric monster.

Nor are this book's front-cover monsters the only ones that will be encountered within the pages of this book. Drawn from an eclectic intermingling of my brand-new writings, updated posts from my *ShukerNature* blog, and expanded versions of articles of mine previously published only in various British magazines and therefore not readily accessible to my very sizeable readership audience outside the UK, so too will be a host of other truly monstrous life-forms. These include such fearsome fauna as still-undiscovered gigantic lizards and giant frogs, a fire-breathing pig reputedly guarding Swedish churchyards, a couple of hitherto little-known but fascinating man-beasts extracted from traditional Chinese and Basuto lore, the much-dreaded were-worms of Tolkien, an extraordinary collection of tangle-tailed rat kings, umbilically-united cat kings, and the astonishing vegetable lamb of Tartary, rodents with horns, devil's corkscrews, and terrible snails, the missing mystery tusk of Rothschild and some putative Nandi bears held captive in England, a surfeit of sea serpents and some very strange sea monster carcasses, the ghastly winged tomb spider of Italy and a golden freshwater oarfish from the Vietnam War, a pack of urban hyaena-dogs on the prowl in New York City, colossal blue elephant-ingesting eels in the Ganges, and much more, too.

So, if you're looking for monsters, you've certainly come to the right place, and opened the right book. Just pray that you don't live to regret your bravery—or foolishness—in having done so. In fact, just pray that you do live...

> Culture teems with animals who have no exact equivalent in nature. A huge, make-believe fauna of monsters, prodigies, and wonders slithers and swarms and storms through all of the arts, as though the natural world were somehow deficient. It is necessary to ask: To what end are these creatures of the imagination? Are they really substitutes for ordinary animals or do they have their own purposes? What are they, where do they come from, and what are they doing here?
> — Paul Shepard, *The Others: How Animals Made Us Human*

What, where, and what indeed? Time to step through the doors of your own perception and read on, as I examine all of the usual—and unusual—suspects in my search for answers to these and many other questions.

Chapter 1:
CAPTIVE NANDI BEARS IN BRITAIN—OR A MORE SLOTHFUL SURPRISE?

> *Periodically come reports from the Kakamega forests*
> *in Kenya of sightings of the Nandi bear. The beast is*
> *described as having a gorilla-like stance with forelimbs*
> *longer than the hind, with clawed feet like a bear and*
> *with a horse-like face. Could the beast be a survivor of*
> *the chalicothere, thought to have become extinct in East*
> *Africa during the Pleistocene? The description above*
> *would fit with the skeletal remains of these extraordinary*
> *animals.*
>
> — R.J.G. Savage and M.R. Long, *Mammal Evolution*

One of the most formidable, ferocious mystery beasts on record, the Nandi bear of Kenya's Nandi and neighboring Kakamega forest regions, was once widely reported, but nowadays it seems to have gone out of fashion—or even out of existence—because there do not appear to have been any documented sightings of it for many years.

As discussed by veteran cryptozoologist Bernard Heuvelmans in *On the Track of Unknown Animals* (1958) and further assessed in my own book *In Search of Prehistoric Survivors* (1995), the Nandi bear seems to have been many things to many people, inasmuch as it was apparently a composite creature, i.e. "created" from the erroneous lumping together of reports describing several taxonomically discrete animals. Some of these are already known to science, but others may not be, at least in the living state.

They include: old all-black ratels (honey badgers) *Mellivora capensis*; some form of extra-savage giant baboon; erythristic (freakishly red-furred) spotted hyaenas *Crocuta crocuta* and/or a supposedly long-extinct lion-sized relative called the short-faced hyaena *Pachycrocuta brevirostris*; the aardvark *Orycteropus afer*; perhaps even a relict true bear related to (or synonymous with) the Atlas bear *Ursus arctos crowtheri*, which existed in North Africa until as recently as the 1870s; and, most fascinating of all, a putative surviving species of chalicothere. The latter were bizarre perissodactyl (odd-toed) ungulates that possessed claws instead of hooves, and which may have been somewhat hyaena-like in superficial appearance (due to their rearward-sloping back) but were much larger in size. According to the fossil record, chalicotheres lingered on until at least as recently as one million years ago in Africa, but died out earlier elsewhere in the world.

The prospect of a modern-day chalicothere being responsible for certain Nandi bear reports was popularized by Heuvelmans, but in spite of common assumption to the contrary, he definitely did not originate this notion. Instead, it was presented and discussed at length as far back as 1931, by Captain Charles R.S. Pitman in the first of his two autobiographical works, *A Game Warden Among His Charges*. Moreover, it was briefly alluded to even earlier, by Charles W. Andrews in his *Nature* article from 1923 regarding the finding of chalicothere fossils in Central Africa. Even the renowned Kenyan palaeoanthropologist Louis S.B. Leakey contemplated it in an *Illustrated London News* article of November 2, 1935. Certainly, the idea has long held a particular fascination for me, because it alone could provide a reasonable explanation why the Nandi bear has seemingly vanished.

Artiodactyls (even-toed ungulates, e.g. cattle, antelopes, giraffes, pigs) were devastated by an epidemic of rinderpest (a morbillivirus) that swept across southern Africa during the late 19th century. In 1995, it was revealed that a distantly related morbillivirus was comparably deleterious to horses (which, like chalicotheres, are perissodactyls). So could a morbillivirus have wiped out a chalicotherian Nandi bear? None of the other Nandi bear identities would be affected by such a disease, so if only these identities were components of the Nandi bear composite (i.e. with no ungulate component ever involved), we would expect Nandi bear reports to be still surfacing, whereas in reality none has emerged for years.

Someone else who was very intrigued by the concept of a chalicotherian Nandi bear was British author and wildlife educator Clinton Keeling, whose death in 2007 robbed the international zoological community of a uniquely knowledgeable expert on the histories and exhibits of zoological gardens, circuses, and menageries (travelling and stationary) throughout Britain and overseas, both in the present and in the past. During the course of a long, productive life as a zoo curator and also travelling widely to schools with animals to entertain and educate generations of children concerning the wonders of wildlife, Clinton wrote and self-published over 30 books (but all of which, tragically, are fiendishly difficult to track down nowadays) documenting wild animal husbandry and also the histories of demised and long-forgotten animal collections. These are a veritable treasure trove of extraordinary information and insights that are very unlikely to be found elsewhere, providing details of some truly remarkable and sometimes highly mysterious creatures that were at one time or another on display in Britain—and which in Clinton's opinion may have included at least three living chalicotherian Nandi bears!

Reconstruction of a chalicothere in life (Hodari Nundu)

Frustratingly, however, I have never managed to obtain a copy of any of Clinton's books. So after he published a summary of his Nandi bear accounts from two of them in the form of a short article appearing within the July 1995 issue of the Centre for Fortean Zoology's periodical *Animals and Men*, I wrote to him requesting further information. In response, he kindly wrote me a very detailed letter, dated July 3, 1995, documenting all that he knew about this extremely exciting possibility and also regarding various other cryptozoological subjects. Its contents made enthralling, thought-provoking reading, but I have never published in any of my books its Nandi bear section (or even any excerpts from it)—until now. So here, for the very first time in book form, is Clinton Keeling's full and thoroughly fascinating account of that tantalizing bygone trio of unidentified captive beasts in Britain that just may have been living Nandi bears:

> Rest assured I shall be happy to assist you in any way possible concerning the "Nandi Bear", of which I am convinced at least three specimens have been exhibited in this country—although their owners had no idea what they were...
>
> I think it would be best if I were to quote directly from two of my books...in this way you'll know as much as I do when you've

finished reading. The following—I'll call it NB1 [i.e. Nandi Bear Case #1]—is from my book *Where the Crane Danced*, written in 1983; I'm dealing with the earliest travelling menageries:

"The first one I have been able to learn anything about must have been operating in the 1730s, and although not even its name has been recorded I was absolutely thrilled to discover that it contained what might well have been proof that an animal that most people relegate to the Loch Ness Monster bin really did exist—and comparatively recently too. In a nutshell, I have always been interested in the mysterious creature usually referred to as the Nandi Bear, which might still exist on the Uashin Gishu Plateau in Kenya; some people swear it was/is a belated Chalicotherium, a primitive ungulate with claw-like hooves which officially became extinct long ago, while others pooh-pooh the whole tale as an utter fabrication. Those who claim to have seen it, though, and they are many, all talk of a Hyena-like creature with the head of a Bear [some descriptions, however, offer the converse description, i.e. hyaena-headed and bear-bodied]. And please note this menagerie that might have shown one was operating getting on for two centuries before Kenya was opened up by Europeans, so in other words no-one had heard of it then. I first came upon this intriguing possibility when looking through some old numbers of *Animal and Zoo Magazine*, the long-defunct publication I mentioned in *Where the Lion Trod* [another of Clinton's books]. In the edition for February 1938 it stated that a reader in Yorkshire had found a bill "two hundred years old" that read:

> "Posted at the sign of the Spread Eagle, Halifax. This is to give notice, to all Gentlemen, Ladies and others, that there is to be seen at the sign of the Coffee House, a curious collection of living creatures..."

"It then went on to list its attractions, chiefly Monkeys and smallish carnivores, the last of which was:

> "A young HALF and HALF; the head of a Hyena, the hind part like a Frieseland [Polar? [this query was inserted by Clinton]] Bear."

"Now it would certainly not have been a Hyena, or a Bear, as clearly whoever penned the advertisement apparently knew what they looked like, so one is left to ponder on this curiosity, which sounds so much like descriptions of that weird threshold-of-science creature which has so often been seen by sober people of high reputation as it has gone slinking through the long grass in the African night."

NB2 [Nandi Bear Case #2] comes in my *Where the Macaw Preened* (1993), and its source is interesting. In *Where the Crane Danced* I dealt in some detail with Mander's Menagerie, a huge display second in size only to Bostock and Wombwell's, and which finally came off the road in 1875. As a result of this, I was contacted by a Mrs Rosanne Eccleston of Telford, Shropshire, who is a descendant of the Manders. She sent me a facsimile of an extremely lengthy advert, placed in a York newspaper in November 1869 which was, in effect, a stocklist of the show at that time (it included such unexpected items as Ligers); Mander was [a] very experienced animal-man, but sometimes he got his geographical area of distribution wrong, usually—and this could be significant—when he'd obtained a rare or obscure species (i.e. not what I call a Noah's Ark animal—Lion, Tiger, Bear, etc.) about which he knew little or nothing. Anyway, I quote directly from the end of the section on Mander's Menagerie in *WTMP* [*Where the Macaw Preened*]:

"I've deliberately left what I consider to have been the most remarkable exhibits until the last, so we can savour them for the marvels that I think they *could* have been. Oddly enough, they were one of the few species to be given what's clearly the wrong area of distribution.

"Listed as 'Indian Prairie Fiends' they were described as:

> Most wonderful creatures. Head like the Hippopotamus. Body like a Bear. Claws similar to the Tiger, and ears similar to a Horse.

"That's all, and forget the inference to North America [i.e. the prairie portion of the name applied to these creatures in the listing], as there's nothing in that part of the world that has ever resembled anything like this, but, descriptions given by Africans apart, this is the best word-picture of the Chimiset or Nandi Bear I've ever happened upon.

"Many people, I know, relegate this astonishing creature to the same category as the Loch Ness Monster and other twilight beasts which might or might not exist, but here I feel they are being unjust as the question should really be "does it *still* exist?", as of all the "mystery" animals this is the one scientific sceptics come nearest to accepting, as paleontologists have learned a great deal about the Chalicotherium—which is believed to be the origin of the Nandi Bear. In short, it resembled a nightmarish (no pun intended) Horse—

in fact it was related to the Equines—which had huge claws and preyed upon other animals, in fact many Africans have stated how fierce it is, and how destructive to their livestock ("Fiends", I trust you've noticed; the only implication so far of viciousness—again, it fits). Readers of *WTCD* [*Where the Crane Danced*] will recall my suggestion that a menagerie touring northern England in the 1730s also boasted a young specimen—which is at least perfectly possible, as there now seems little doubt that a small relict population of Chalicotheriums (Chalicotheria?) hung out on the Uashin Gishu Plateau in East Africa until the very end of the 19th century, when it was wiped out by the great rinderpest epidemic of 1899. Remember, it *was* an ungulate, despite not having hooves and eating flesh. What a pity Mr Mander didn't think anyone would be interested to learn what he fed his specimens on!"

All of which brings up some fascinating points. For a start, on the face of it, it sticks out a mile that the two reports are of completely different animals, but whereas the "Halifax" creature was a classic description of the beast seen so often in Africa a century ago, the "York" one is a word-perfect reconstruction of modern assessments of what the chalicotherium must have looked like—even to the Horse-like (Hippopotamus) head and massive claws. I agree it sounds paradoxical, but here are good descriptions of the creatures seen in the field by traveller and tribesman, and the armchair explorers' and scientists' word-picture of what it must have resembled. In other words, there's a strong case for each.

An extremely impressive brief can be made for Mander's animals, as it's the only species in his list with a "made-up" name; all others either have appellations still in use, or old but then perfectly acceptable ones, such as "Yaxtruss" for Yak and "Horned Horse" for Wildebeests: this one alone has an outlandish name. It's very highly significant, too, that again it's the only one to be described in detail— presumably on the assumption that most people would know what a Camel or a Zebra or a Kangaroo was. In other words Mander, who most certainly knew an extremely wide range of species, hadn't the slightest idea of what the Indian Prairie Fiends really were.

I cannot emphasise strongly enough that whatever these animals were, they would certainly have been on show, and more or less as described, as contrary to popular belief, the showmen of yesterday might have exaggerated the size or physical attributes of their exhibits, but they certainly didn't advertise what they hadn't got. They were not fools, and knew full well the measures a mob of 19th century colliers, artisans, idlers and toughs would take if it thought it was being swindled or "conned".

Most unfortunately it didn't enter the heads of these very materialistic travellers to keep Occurrences Books (other than places visited and money taken) so unfortunately we'll probably never know how these I.P.F.s [Indian Prairie Fiends] were obtained, how many there were, their diet, how long they lived, or—very important— what became of them. I mention this because there was often an arrangement with museums whereby unusual cadavers were eagerly purchased (in Weston Park Museum, Sheffield, for example, there are two hybrid big Cat cubs purchased long ago from a travelling show) so I suppose it's *just* possible, in some dusty storeroom, there could be a couple of interesting skulls or pelts.

A fascinating and very thought-provoking communication, to say the least! However, it contains certain assumptions that need to be addressed and rectified.

First and foremost: contrary, to Clinton's claims, the chalicotheres were not carnivorous, they were wholly herbivorous—a major conflict with the Nandi bear's bloodthirsty rapaciousness that Heuvelmans sought to explain by speculating that perhaps the occasional sight of so extraordinary a beast as a chalicothere, armed with its huge claws, was sufficient for a native observer to assume (wrongly) that they had spied a bona fide Nandi bear. In other words, even if there are any living chalicotheres, these perissodactyl ungulates are only Nandi bears by proxy. Having said that, however, as I pointed out in my book *In Search of Prehistoric Survivors*, certain other perissodactyls, such as some zebras, tapirs, and most notably the rhinoceroses, can be notoriously bellicose if confronted. If the same were true of chalicotheres, one of these horse-sized creatures with formidable claws and an even more formidable, highly aggressive defensive stance would definitely make a veritable Nandi bear, even though it wouldn't devour its victim afterwards.

When referring to the Halifax mystery beast (NB1), Clinton wonders whether the "Frieseland [sic] bear" that it was likened to was a polar bear. In reality, however, the only bears native to Friesland, which is part of the present-day Netherlands, are brown bears *Ursus arctos*. Consequently, this suggests that the animal's hind parts resembled a brown bear's, not a polar bear's.

My greatest concern, however, is Clinton's determination to believe that the Halifax mystery beast and the York mystery beasts (NB2) were the same species (even after stating himself that at least on first sight the two reports describe two totally different types of animal). Personally, I fail to see how a hyaena-headed creature can be one and the same as a hippo-headed creature—unless, perhaps, these were simply differing ways of emphasizing that the creatures had big, noticeable teeth? In the same way, likening their ears to those of horses might indicate that, as with horses' ears, theirs were noticeable without being prominent. Alternatively (or additionally?), describing an animal's head as hippo-like may imply that it had large, broad nostrils and/or mouth.

Clinton's statement that the hippo-headed York cryptids corresponded with a chalicothere's appearance cannot be countenanced, because chalicotheres' heads were horse-like (which hippos aren't), and chalicotheres didn't have big teeth. So even if the hippo-head comparison was just an allusion to the size of the York cryptids' teeth, a chalicothere identity is still ruled out for them.

My own view is that if either of the two cryptid types documented here were a Nandi bear, it is more likely to have been the hyaena-headed, bear-bodied Halifax animal. Even so, this latter beast sounds very reminiscent of a scientifically recognized but publicly little-known species whose distinctive appearance would certainly have made it a most eye-catching exhibit. Today, three species of true hyaena exist, two of which—the striped hyaena *Hyaena hyaena* and the earlier-mentioned spotted hyaena—are familiar to zoologists and laymen alike. The third, and rarest, conversely, is seldom seen in captivity and is elusive even in its native southern African homeland.

This reclusive species is the brown hyaena *H. brunnea*, which just so happens to combine a hyaena's head with a dark brown shaggy-furred body that is definitely ursine in superficial appearance (as I can personally testify, having been fortunate enough to espy this species in the wild in South Africa), and especially so in the eyes of a zoologically untrained observer. So could the Halifax mystery beast have been a sub-adult brown hyaena, captured alive alongside various more common African species and then transported to Britain with them, where it was destined to be displayed to a wide-eyed public that had never before seen this exotic-looking species? It is certainly not beyond the realms of possibility and is a more plausible identity than a Nandi bear.

As for the York cryptids, an identity very different from that of a Nandi bear but equally cryptozoological in nature came to mind as soon as I first read Clinton's account of them.

Clinton discounted their "Indian prairie fiend" name by accurately stating that nothing resembling them is known from North America. But what if they had come from South America instead? The "Indian" reference could simply have been to whichever native Indian tribe(s) shared their specific distribution in South America. And could it be that "prairie" was nothing more than an alternative name for "pampas," perhaps substituted deliberately by Mander as he knew

Early, vintage photograph of a brown hyaena in captivity (public domain)

that "prairie" would be a more familiar term than "pampas" to his exhibition's visitors?

But does the South American pampas harbor a creature resembling those cryptids from Halifax? Until at least as recently as the close of the Pleistocene epoch a mere 10,000 years or so ago, this vast region (encompassing southernmost Brazil, much of Uruguay, and part of Argentina) did indeed harbor large shaggy bear-like beasts with huge claws, noticeable ears, plus sizeable nostrils and mouth. I refer of course to the ground sloths—those burly, predominantly terrestrial relatives of today's much smaller tree sloths. Moreover, the pampas has hosted several modern-day sightings of cryptids bearing more than a passing resemblance to ground sloths—and thence to the Halifax mystery beasts.

Some species of ground sloth were truly gigantic, but others were of much more modest proportions, and there is no doubt that a medium-sized species of surviving ground sloth would solve a number of currently unresolved cryptozoological conundra, not least of which is the identity of the mystifying Halifax beasts. Specimens of many other South American beasts were commonly transported from their sultry homelands and exhibited in Europe back in the days of travelling menageries here. Could these have included a couple of ground sloths? In addition, armed with such huge claws, a cornered ground sloth might well be more than sufficiently belligerent if threatened or attacked to warrant being dubbed a fiend.

So, who knows—perhaps the hypothetical dusty museum storeroom postulated by Clinton as a repository for some mortal remains of the Nandi bear may contain some modern-day ground sloth cadavers instead? It certainly wouldn't be the first time that surprising and highly significant zoological discoveries have been made not in the field but within hitherto unstudied or overlooked collections of museum specimens.

Reconstruction of ground sloth in life (public domain)

Chapter 2:
BILBO BAGGINS VERSUS THE MONGOLIAN DEATH WORM?

> *In the tales of the Hobbit folk there lived in the Last Desert, in the East of Middle-earth, a race that was named the Wereworms. Though no tale of the Third Age of Sun tells of these beings, the Wereworms were likened to Dragons and serpents. To Hobbits they were perhaps but memories of those creatures that stalked the Earth during the Wars of Beleriand in the First Age.*
>
> — David Day, *A Tolkien Bestiary*

The works of J.R.R. Tolkien contain a number of creatures with some pertinence to cryptozoology, such as giant spiders, dragons, and the dreaded watcher in the water (a monstrous freshwater cephalopod?). Yet perhaps the most unexpected as well as the most fascinating Tolkien reference to a cryptid, which occurs in *The Hobbit* (1937), is so brief and inconspicuous that it can be easily passed by or even entirely overlooked, with its cryptozoological significance not even registering upon the reader. This is a great tragedy, because, remarkable as it may seem, the mystery beast in question is none other than the extraordinary Mongolian death worm!

In "An Unexpected Party," which is the opening chapter of *The Hobbit*, the hobbit in question, Bilbo Baggins, has received an unheralded visit at his home, Bag End, from the wizard Gandalf and a company of dwarves, whom Gandalf has informed would do well to include Bilbo, as a burglar, in their planned quest to retrieve their stolen gold from the great dragon Smaug. The dwarves, however, are far from convinced that Bilbo would serve well in this capacity, and they air their doubts very vocally in Bilbo's parlor while he is in the drawing-room (but, unbeknownst to them, still within ear-shot of their protestations). Angered by their dismissive attitude, Bilbo strides back into the parlor and boldly proclaims that he is more than capable of fulfilling the role that they wish him to undertake:

> Tell me what you want done, and I will try it, if I have to walk from here to the East of East and fight the wild Were-worms in the Last Desert.

Brave words indeed, but also very puzzling ones, because they are never explained nor even referred to ever again either in this or in any other Tolkien novel. They are presumably said by Bilbo in a figurative sense, to convey that he is willing to tackle anything, however perilous. But even so, what exactly *are* the wild Were-worms that they refer to, and where is the Last Desert?

The term "worm" has a number of different animal-related meanings. Its most familiar zoological meaning is as a contraction of "earthworm"—the common name for most terrestrial oligochaetes. However, many other, unrelated zoological invertebrate taxa that include long, elongate species also have "worm" in their names—tapeworms, peanut worms, acorn worms, beardworms, thorny-headed worms, ragworms, roundworms, flatworms, etc. There are even a few worm-dubbed vertebrates, such as the slow worm *Anguis fragilis* (a species of limbless lizard).

In zoomythology, "worm" is one of several related terms—others include "orm," "ormer," and "wyrm"—applied to certain serpent dragons (i.e. limbless, wingless dragons that basically resemble huge serpents except for their dragon-like head), such as Britain's Lambton worm, Linton worm, and Kellington worm. This category of dragon was also often characterized by noxious breath (rather than breathing fire), and the ability to rejoin into a single entity again if cut up into segments.

Bilbo Baggins as personally interpreted by Swedish artist Richard Svensson (Richard Svensson)

In Tolkien's works, conversely, he applies the term "worm" to a very elongate-bodied form of classical dragon, i.e. equipped with four legs, a pair of wings, and the ability to breathe fire. Smaug in *The Hobbit* was a prime example of Tolkien's dracontological definition of "worm."

Consequently, the were-worm may be a bona fide type of invertebrate worm, albeit one of formidable size and/or temperament if it warranted being fought against (as opposed merely to being trodden upon!); or it could be a Smaug-like dragon. But what about the "were" component of its name?

This prefix, from the Old English *wer*, generally denotes "human"—

hence a werewolf, for instance, is a human that can transform itself into a wolf, a weretiger is a human that can transform itself into a tiger, and so on. Does this mean, therefore, that a were-worm is a human that can transform itself either into a gigantic invertebrate-type worm or, perhaps more plausibly, into a dragon? Alternatively, is it a true invertebrate-type worm, or a true dragon, but one that exhibits highly advanced, human-like intelligence? Or could it even refer to a being that was half-human, half-dragon, akin in form perhaps to the ancient Indian snake deities or nagas, which possessed human heads (and sometimes thorax and arms too) but serpent bodies? Any of these solutions, however, would involve an entity of truly monstrous nature.

As for the Last Desert, where these were-worms reputedly dwell: all that appears to be known about this arid realm is that according to hobbit folklore, it is located at the very easternmost end of Middle-earth, and therefore lies far to the east of the Shire where the hobbits live.

But where does the Mongolian death worm fit into all of this? As any self-respecting cryptozoologist will know, this much-dreaded cryptid allegedly inhabits the Gobi Desert, and according to the nomads living in fear of it there, it can not only squirt a lethal acidic venom at anyone confronting it, but also kill directly via touch (or even indirectly if a person touches it with an implement made of metal) in a mysterious manner that is extraordinarily reminiscent of electrocution.

Yet even the versatile death worm cannot transform into a human or vice-versa. So what connection can there be between this cryptid and Tolkien's were-worm, other than that they both inhabit deserts?

A representation of the Mongolian death worm (Thomas Finley)

A comprehensive two-part study of *The Hobbit*, entitled *The History of The Hobbit*, was published in 2007 by HarperCollins in the UK (and by Houghton Mifflin in the USA), containing Tolkien's unpublished drafts of *The Hobbit*, together with commentary written by Tolkien scholar John D. Rateliff. These drafts revealed how this novel had undergone many changes, some minor, some major, between the very first version and the final, published edition. One such change is of paramount important to the subject of this present chapter because it concerns Bilbo's statement regarding the were-worms.

It turns out that in the very first, original draft of *The Hobbit*, that statement made no mention at all of were-worms, or of the Last Desert. Instead, what it did state, very thought-provokingly, is as follows:

> [that Bilbo would walk to] the Great Desert of Gobi and fight
> the Wild Wire worms of the Chinese.

How remarkable that the Last Desert as named in the final, published edition of *The Hobbit* was clearly inspired, therefore, by none other than the real-life Gobi Desert. And no less significant is that the ostensibly shape-shifting were-worm was apparently no such thing in the original draft of *The Hobbit*, being a wire worm instead. But what did this term signify?

In view of the reference to the Chinese, could it have referred to one of those famously serpentine-bodied Oriental dragons? However, they tend to waft languorously through the skies, or rise up from the seas or from deep freshwater pools, rather than reside in deserts, and are often viewed in ancient Eastern traditions as deities. So this identity for Tolkien's wild wire worms seems somewhat unlikely. And why, in any case, would he have applied the adjective "wire'" to such dragons?

Certainly, the term "wire worm" is intriguing, inasmuch as in zoological parlance a wire worm is the elongate limbless worm-like larva of a click beetle,

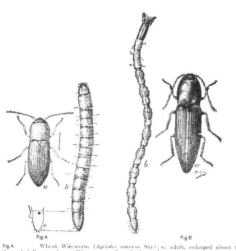

Fig A   Wheat Wireworm (*Agriotes mancus* Say); *a*, adult, enlarged about five times; *b*, full-grown larva (Wireworm), enlarged about three times; *c*, side view of last segment of larva. (*From U. S. D. A. Bull.* 156.)
Fig B   Corn and Cotton Wireworm (*Horistonotus uhleri* Horn); *a*, adult, enlarged about ten times; *b*, full-grown larva (Wireworm), enlarged over four times. (*From U. S. D. A. Bull.* 156.)

Two species of click beetle and their respective wire worm larval form (public domain)

belonging to the family Elateridae. But I hardly think that Tolkien was referring to some hobbit-inimical, hyper-aggressive click beetle grub when writing of "Wild Wire worms."

All of which brings us, therefore, to the Mongolian death worm. This cryptozoological creature has definitely—indeed, exclusively—been reported from the Gobi Desert, and is, undeniably, zoologically worm-like in overall appearance. Indeed, its local names, "allergorhai-horhai" and "allghoi-khorkhoi," both translate as "intestine worm," as it is likened by the nomads to a worm that resembles an animate intestine. But is it conceivable that Tolkien had heard of such an entity, especially way back in the 1930s? After all, Western cryptozoology itself did not become aware of it until the 1990s, when Czech explorer Ivan Mackerle began searching for and writing about the Gobi's reputed vermiform inhabitant after having researched its history in Russian and Mongolian documents.

Nevertheless, Tolkien, as a highly erudite, eclectic reader, may indeed have known of such a beast, thanks to the publication in 1926 of *On the Trail of Ancient Man*, written by eminent American palaeontologist Roy Chapman Andrews. This bestselling book concerns the American Museum of Natural History's famous Central Asiatic Expedition of 1922 to the Gobi, led by Andrews, in search of dinosaur fossils, but it also includes a mention of the Mongolian death worm—which as far as I am aware is the earliest such mention of it in any Western publication.

In order to obtain the necessary permits for the expedition to venture forth into the Gobi, Andrews needed to meet the Mongolian Cabinet at the Foreign Office. When he arrived, he discovered that numerous officials were attending their meeting, including the Mongolian Premier. After Andrews had signed the required agreement in order to obtain the expedition's permits, the Premier made one final but very unusual and totally unexpected request:

> Then the Premier asked that, if it were possible, I should capture for the Mongolian government a specimen of the *allergorhai-horhai*. I doubt whether any of my scientific readers can identify this animal. I could, because I had heard of it often. None of those present ever had seen the creature, but they all firmly believed in its existence and described it minutely. It is shaped like a sausage about two feet long, has no head nor legs and is so poisonous that merely to touch it means instant death. It lives in the most desolate parts of the Gobi Desert, whither we were going. To the Mongols it seems to be what the dragon is to the Chinese. The Premier said that, although he had never seen it himself, he knew a man who had and had lived to tell

the tale. Then a Cabinet Minister stated that "the cousin of his late wife's sister" had also seen it. I promised to produce the *allergorhai-horhai* if we chanced to cross its path, and explained how it could be seized by means of long steel collecting forceps; moreover, I could wear dark glasses, so that the disastrous effects of even looking at so poisonous a creature would be neutralized. The meeting adjourned with the best of feeling.

Call me a cynic, but I have the distinct impression that Prof. Andrews did not take the death worm too seriously. In any event, he certainly didn't succeed in finding one, which is probably no bad thing—bearing in mind that he had planned to pick up with steel forceps a creature that had allegedly brought about instantaneous death to a fellow geologist who had prodded it with a metal rod!

During the 1920s, the American Museum of Natural History sent forth several additional Central Asiatic Expeditions to Mongolia and China, and in 1932 a major work, *The New Conquest of Central Asia*, was published, documenting all of them, with Prof. Andrews as its principal author. The first volume in the series *Natural History of Central Asia* (edited by Chester A. Reeds), it contained a brief section entitled "The Allergorhai Horhai":

> At the Cabinet meeting the Premier asked that I should capture for the Mongolian Government a specimen of the *Allergorhai horhai*. This is probably an entirely mythical animal, but it may have some little basis in fact, for every northern Mongol firmly believes in it and will give essentially the same description. It is said to be about two feet long, the body shaped like a sausage, and to have no head or legs; it is so poisonous that even to touch it means instant death. It is reported to live in the most arid, sandy regions of the western Gobi. What reptile can have furnished the basis for the description is a mystery!
>
> I have never yet found a Mongol who was willing to admit that he had actually seen it himself, although dozens say they know men who have. Moreover, whenever we went to a region which was said to be a favorite habitat of the beast, the Mongols at that particular spot said that it could be found in abundance a few miles away. Were not the belief in its existence so firm and general, I would dismiss it as a myth. I report it here with the hope that future explorers of the Gobi may have better success than we had in running to earth the *Allergorhai horhai*.

If Tolkien had read either or both of these books, and as someone passionately interested in archaeology it is by no means an unlikely possibility, then he would indeed have learnt of the dreaded death worm, whose sensational nature might very well have impressed him sufficiently to incorporate a version of this creature in his first draft of *The Hobbit*. Moreover, Andrews's comparison of the death worm's significance to the Mongolian people with that of the dragon to the Chinese may even have inspired Tolkien's otherwise-opaque linking of the wild wire worms to the Chinese.

As for why the wild wire worms were replaced in later drafts by wild were-worms, and the Gobi Desert replaced by the Last Desert, who can say? Perhaps Tolkien felt that the latter versions were more compatible with the entirely fictitious Middle-earth than were a real desert and a semi(?)-mythical creature from Mongolian tradition referenced to in a real scientific publication.

Wood carving of a death worm-like creature in Gobi museum near Dalanzadgad, Mongolia (Ivan Mackerle)

Of course, this is all very speculative, as there is no firm evidence that Tolkien ever did read or even know of Andrews's books, but that memorable sentence in Tolkien's original draft of *The Hobbit* remains a compelling enigma. And if nothing else, it conjures up the truly surreal scenario of a hobbit doing battle with the Mongolian death worm—which is surely worthy of a novel in its own right!

Chapter 3:
MONITORING SOME LESSER-KNOWN GIANT MYSTERY LIZARDS

It is interesting to recall that many years ago a visitor to this region [St. Lucia estuary on the coast of Zululand, South Africa] stated that he had seen a dragon-like creature on the side of St. Lucia. The creature was watched for several minutes through a pair of powerful binoculars. It was estimated to be about 9 feet in length and to weigh about 250 lbs. It was a slatey colour with small scales, with a claw protruding several inches beyond the others. The creature, in fact, had the general appearance of a lizard.
— W.L. Speight, "Mystery Monsters in Africa,"
Empire Review (1940)

Many new and sometimes quite striking species of varanid (monitor lizard) have been discovered lately. Judging from the chronicles of cryptozoology, however, there may be a number of far more spectacular forms still awaiting formal scientific unveiling. Some of these cryptozoological mega-lizards, such as New Guinea's artrellia, South Africa's das-adder, and Australia's mungoon-galli and burrunjor (or alleged mega-lizards, in the highly controversial case of Assam's buru), have been extensively documented elsewhere by others and myself. However, several much less familiar (yet no less fascinating) examples have received far less publicity, even within cryptozoological circles—until now.

MEGA-MONITORS FROM THE CONGO?
There are some particularly intriguing examples of putative giant varanids on record from Africa.

The Likouala swamplands in the People's Republic of the Congo (formerly the French Congo) have long attracted cryptozoological attention, thanks to the claimed existence in these enormous, scarcely-explored wildernesses of a reclusive amphibious creature known as the mokele-mbembe—whose long neck and tail, sturdy body, and four powerful limbs confer upon it more than a passing resemblance to a sauropod dinosaur! During the 1980s, Roy P. Mackal—then a Chicago University biochemist and enthusiastic spare-time cryptozoologist—

led two expeditions to this locality in search of its mysterious "water dragon." And although the team failed to spy it, they did obtain considerable new native testimony concerning not only this famous cryptid but also a number of other, less familiar Congolese ones.

One of the most intriguing of these is the nguma-moneme. This is another amphibious form, equally adept moving in water and on land, and is described by the native pygmies as being an extremely elongate and low-slung reptilian beast, almost like a four-limbed snake and measuring up to 33 feet in total length, but with a saw-like dorsal ridge or frill of small triangular projections running down the total extent of its back and its lengthy tail. It also has a forked tongue.

When documenting this distinctive mystery animal in his book *A Living Dinosaur?* (1987), Mackal opined that it could well be an unknown, primitive species of semi-aquatic monitor lizard, directly derived from ancestral rather than from more advanced varanids. Interestingly, the Nile monitor *Varanus niloticus* is ubiquitous in the Congo basin, but even if gigantic specimens unseen by scientists exist here, this species still lacks the serrated dorsal ridge of the nguma-monene.

Reconstruction of the likely appearance of the nguma-monene (Roy P. Mackal)

Critics of the "prehistoric survivor" sauropod identity supported by some cryptozoologists for the mokele-mbembe have suggested that this too may be a monitor lizard, albeit a huge and very specialized one. Eyewitness reports claim that the mokele-mbembe is up to 30 feet long, reddish-brown, and strictly herbivorous, reaching up with its long neck to consume *Landolphia*—a tropical creeper with large round or pear-shaped fruit and white blossoms.

Although monitors do have discernible necks, no known species has one as lengthy as the mokele-mbembe's *Diplodocus*-like version. Moreover, hardly any of the many known species of living monitor are herbivorous. Nor are there monitors from the fossil record that compare with the mokele-mbembe. And as such lizards are common in the Likouala swamps, if this is what the mokele-mbembe is too, why do the natives not identify it as such? Instead, when they were shown series of animal images by Mackal's team and also by researchers from other visiting expeditions, they consistently selected pictures of sauropods, not monitors, as offering the closest visual match to it.

THE DRAGONS OF BABYLON

The mokele-mbembe has also been linked to another, geographically far-removed mystery beast—the Babylonian mushussu or sirrush. Along with the lion and the bull, the mushussu was one of the three regal beasts depicted in profile on Babylon's magnificent Ishtar Gate, erected during the reign of King Nebuchadnezzar II (605-562 BC) and dedicated to Ishtar, the Babylonian fertility goddess.

Richly adorned with glazed bricks of dazzling cobalt blue and several horizontal rows of these three animal types depicted in realistic form with golden bricks, this arched gateway was a veritable wonder, spanning the processional way between the temples of Ishtar and the sun god Marduk. After the fall of Babylon in c.39 BC, however, it remained buried in the sands for many centuries, until finally rediscovered by German archaeologist Robert Koldewey during the late 19th century. Excavations began in 1899 and took several years to complete. It can now be seen in all of its fully restored resplendence within the Vorderasiatisches Museum in Berlin, Germany, and having viewed it there back in 1983 I can confirm that it is indeed magnificent.

The mushussu had a long slender scaly body and tail, which was carried almost vertically. By way of symmetry, it also had a long and near-vertical neck, with a typical dragon's head bearing a pair of curled horn-like structures and what may be a single vertical horn upon its brow like a reptilian unicorn. A long forked tongue emerged from its mouth. Its four legs were all sturdy, but the forelimbs were quite different from the hind ones. Unscaled, they terminated in mammal-like clawed paws; whereas the hind limbs' upper regions were scaled, and their lower regions resembled those of a mighty eagle, armed with huge talons.

Of interest is whether this dragon was of the same type as the example referred to in "Bel and the Dragon"—one of the books of the biblical Apocrypha. This book tells of how the people in Babylon once worshipped as a living deity a dragon that lived in the temple of one of their gods, Bel—until it was choked to death by Daniel, to prove that it was mortal, just like themselves, and therefore not a genuine god deserving of their veneration.

Line drawing of the mushussu as portrayed upon the Ishtar Gate of Babylon (public domain)

Some cryptozoologists have speculated whether the mushussu was actually based upon distorted travelers' tales of the mokele-mbembe, which has been reported not only from the Congo but also from Cameroon, Gabon, and elsewhere in Central Africa under a variety of different local names but of identical morphological form. In addition, there have even been bold suggestions that perhaps Bel's dragon was itself a young mokele-mbembe that had been brought back to Babylon alive. Certainly, temple carvings and bas-reliefs dating from that period in ancient history provide evidence that a number of exotic animals were indeed transported here from tropical Africa as exhibits or gifts for the rulers of Babylon.

To my mind, however, I think it far more likely that these Babylonian dragons were nothing more than monitor lizards, albeit very sizeable ones. Pertinent to this prospect is a report from 1173 AD by Castillian traveler Benjamin of Tudela. According to him, he had spied many Mesopotamian "dragons," which he also claimed were so infesting the ruins of King Nebuchadnezzar II's palace as to render them inaccessible. These again were undoubtedly large monitors, which are indeed native to this region of the Middle East. One notable example is the desert monitor *Varanus griseus*, which can attain a total length of almost 6 feet; and the widely distributed Bengal monitor *V. bengalensis*, occurring in southeastern Iraq, can reach 5.75 feet. Both of these would certainly appear somewhat dragonesque in form to any non-zoological observer.

Returning to central-west Africa: In 1950, zoologist Jorgen Birket-Smith reported observing a sandy-colored monitor in Cameroon that he estimated to be at least 7 feet long. And in 1984, another zoologist, Toshitaka Iwamoto, was exploring this same country's jungles when he spied a exceedingly large grey monitor that he estimated to be well over 7 feet long, as he subsequently recalled to American varanid specialist Mark K. Bayless.

THE AFA OF IRAQ

The world's largest living species of lizard is an Asian monitor—the Komodo dragon *Varanus komodoensis*, which can grow up to 10 feet long and inhabits Komodo and certain other small islands in Indonesia's Lesser Sundas group. Remarkably, however, this spectacular varanid has only been known to science for little more than a century; it was not formally described and named until 1912. In addition, judging from much more recent reports of huge lizards emanating from elsewhere in Asia, this exotic continent may *still* harbor some truly extraordinary mega-monitors, occasionally encountered but continuing to evade scientific discovery and official recognition.

Take, for example, the afa. One of the most obscure cryptozoological lizards, it was briefly reported by explorer Sir Wilfred Thesiger in his book *The Marsh Arabs* (1964). Also known as the Madan, the Marsh Arabs inhabited the marshlands of the Tigris and Euphrates rivers in the south and east of Iraq, and along the Iranian border—formerly a vast area of wetland covering more than 5.8 square miles. According to Thesiger, who had lived among them intermittently for eight years during the 1950s prior to the Iraqi revolution of 1958, the canoe-borne Madan claimed that the marshes at the mouth of the Tigris in Iraq was home to a monstrous lizard, which they termed the afa.

Little else appears to have been documented concerning it. As various varanid species are native to this region of Asia Minor, however, the afa may well be one too, albeit bigger than those formally recognized by science here—and hence either an unknown giant species, or based upon sightings of extra-large specimens of some known species.

Sadly, however, the afa's taxonomic identity may be nothing more than of academic interest nowadays. This is because following the Gulf War in 1991, the Iraqi government initiated a major program to divert the flow of the Tigris and Euphrates Rivers away from the marshes in retaliation for a failed Shia uprising among the Arabs living there.

This not only eliminated the Madan's food sources, forcing them to move elsewhere, but also turned the marshes themselves into a desert. Consequently, the afa may well have been exterminated, especially if it were primarily aquatic, because I am not aware of any post-1991 reports alluding to it.

A SUMATRAN MEGA-SAURIAN

In 1561, a Portuguese maritime vessel named the *Sao Paulo* was shipwrecked on the coast of Sumatra. One Tuesday evening (April 1st, but fortunately long before April Fool hoaxes had become popular), some of the crew allegedly encountered two huge lizards. One of them swiftly disappeared into the forest with a loud noise, but the other made off towards the sea, and was estimated to be around 16.5 feet long! As big as a barrel in girth, it had green scales on its back and was also handsomely adorned with a series of black bands. When it saw some of the crew approach, however, it rushed at them with its mighty jaws open, which were claimed to have been wide enough to hold a cow in their gape.

The crew promptly fled, taking refuge on top of a series of high blocks of rock, but when the great lizard tried to force its way between two of these, it became firmly wedged, struggling violently, but finding itself unable to pull free. When the crew saw this, they cautiously clambered down from the rocks and swiftly dispatched the trapped lizard using their arquebuses (early muskets) and spears. They then stripped its flesh from its skeleton, which provided more than enough meat for the entire shipwrecked crew to feast upon, and they claimed that when roasted it had the taste and flavor of mutton.

THE SYMBIOTIC JHOOR AND OTHER AMPHIBIOUS EXAMPLES

One of the most intriguing mystery mega-lizards of Asia is surely the Indian subcontinent's jhoor. For according to reports of this mysterious creature, said to resemble a gigantic amphibious monitor up to 20 feet long, it seemingly exists in a mutually beneficial relationship with the estuarine crocodile *Crocodylus porosus*, the two great reptilian species sharing prey and nesting sites, while not attacking one another. It has been reported from a number of Indian locations, including Gujarat State's Gir Forest National Park and the Kathiawar Peninsula, as well as at the mouths of the Ganges in the Sundarbans, which overlap India and (mainly) Bangladesh.

What may well be the same mystery lizard has also been reported from Myanmar (Burma) and Bhutan. Indeed, some of them were once sighted in a northern Bhutanese lake by none other than the King of Bhutan himself, some time slightly prior to the mid-1980s.

In 1996, in a letter published by the magazine *Reptiles*, Chad Steffen from Kentucky, USA, recalled how his stepfather had informed him that while serving in Vietnam during the Vietnam War, he had spied some huge lizards while looking down from his helicopter as it flew along the South China Sea, northeast of Vung Fau in what was then South Vietnam. He said that they were scavenging along the beach while the tide was out, and estimated them to be more than 12 feet long. The only notably large species of monitor known to exist in this locality is the salvator or Asian water monitor *Varanus salvator*, but no specimen attaining the length cited by Steffen's stepfather has ever been verified. Hence it is possible that he overestimated the size of the lizards that he saw. Having said that, however, his military experience makes it likely that his size estimate would have been more accurate than that of the average, untrained observer, thus making this report more significant than might otherwise have been the case.

KOMODO DRAGONS BEYOND KOMODO?

Despite its common name, the Komodo dragon is not confined entirely to Komodo. On the contrary, it has also been confirmed to inhabit the neighboring Lesser Sundas islands of Rintja, Padar, and Flores.

As recently as 1980, moreover, a currently unconfirmed report was filed (and subsequently included by varanid expert Walter Auffenberg in his definitive tome *The Behavioural Ecology of the Komodo Monitor*) that suggested this mightiest of known varanids alive today might also exist on a fifth Lesser Sundas island—Sumbaya.

A Komodo dragon (Karl Shuker)

So the Komodo dragon may still surprise science via its unwonted elusiveness for a beast so big—and aggressive.

THE MO'O—GIANT LIZARD DEITY OF HAWAII

One of the most prevalent yet mysterious creatures in traditional Hawaiian mythology is the mo'o or moho. According to legend, this ancient, monstrous animal, which is invariably associated with water, functions as a guardian spirit deity, variously protecting individuals, families, districts, and places (particularly fishponds), but it can be notoriously capricious. The mo'o resembles a huge shiny-black dragonesque lizard measuring anything from 12 and 30 feet in its true form, but possesses the ability to shape-shift (sometimes appearing as a normal, tiny gecko lizard, other times as a beautiful seductive woman), and communicates its wisdom to humans in dreams. It inhabits deep inland fishponds, where its presence can be confirmed if there is foam upon the water surface. Fishes obtained from such a pond will taste bitter, further proof of the mo'o's presence.

For the most part, mo'os remain hidden, spending their time ecstatically consuming the sacred but intoxicating awa root, which induces them to twist from side to side like a canoe's keel when in water, but they can be seen when the initial flames of a fire light the altars created for them near to their ponds. Interestingly, the petrified head and tail of the mo'o guardian of Puna district on Hawaii are said to be visible at the bottom of two pools—one at Punalua, and the other at Kalapana, half a mile away. Moreover, a female mo'o called Mokuhinia has reputedly made a number of appearances at various western locations on Maui—most notably in 1838, when she allegedly appeared to "hundreds of thousands" of people gathered at her pond.

Three significant mo'os—respectively named Kilioe, Koe, and Milolii—inhabited precipices in the northern coastal region of Kauai. Another two lived in the Wailuku River close to Hilo on Hawaii, and would permit travelers to use them as footbridges to cross the river. Sometimes, however, they would tip the hapless humans into the river and drown them. Consequently, these treacherous monsters were eventually destroyed. According to traditional Hawaiian legend, the implacable enemy of unfriendly mo'os was Hiiaka, goddess of lightning, who slew many such monsters.

The big zoological problem with regard to the mo'o—as big as the mo'os themselves—is where did the Hawaiian belief in such creatures come from, bearing in mind that there is no species of giant reptile native to these islands. True, there are at least eight species of gecko inhabiting them, as well as some anoles and skinks, but these lizards are all very small. Surely, the mighty mo'o was not inspired by the sight of such diminutive, harmless reptiles as these?

Interestingly, legends of giant lizard-like reptiles are common throughout Polynesia (especially Tahiti) and New Zealand. So the original Polynesian colonists of Hawaii probably brought their belief in giant reptiles with them, based perhaps upon ancestral memories of large monitors (or even crocodiles?) back in their original Indonesian homelands, and resulting in the Hawaiian mythology of the mo'o. According to some researchers, Tahitians definitely brought their worship of mo'o deities to Hawaii, as the royal Oropa'a family of Tahiti worshipped them.

Much less probable, but not impossible, is that perhaps Hawaii was home in distant ages past to a species of very large lizard now long-extinct, whose fossil or subfossil remains have yet to be discovered (unless the preserved head and tail of the Puna mo'o constitute such evidence?). There are many Pacific and Indian Ocean islands that were indeed home in the past to such creatures, and sometimes far stranger ones, such as the now-demised horned turtle and land crocodile of New Caledonia. And Hawaii itself was formerly populated by a diverse range of sizeable avifauna, including several species of large flightless ducks and ibises, all now extinct, which such a reptile might have fed upon. As already noted, this is all highly speculative, but it is certainly a mystery how such a detailed, longstanding tradition of giant lizards could have arisen in an archipelago of Pacific islands where such creatures have never been known to exist.

EXPOSING THE INDONESIAN GIGA-GECKO—A HOAX OF MONSTROUS PROPORTIONS!

Finally, although not involving a varanid, no chapter on giant mystery lizards could surely consider itself at an end without having included the following case, if only because it is as entertaining as it is educational.

In recent times, a very striking photograph has attracted appreciable online attention because it ostensibly depicts a gecko of truly gargantuan proportions—not so much a mega-gecko as a veritable giga-gecko! In reality, however, as I swiftly realized when observing it, what this photo truly depicts is something very different from what it may initially seem to do.

After conducting some online research, I was able to trace the photo back to a news article posted in May 2010 by a Rudy Hartono on a website entitled *My Funny* (at: http://funfunpics.blogspot. co.uk/2010/05/giant-gecko-sold-at-least-usd-200000.html), which didn't bode well for the article's contents having a sound scientific basis. That in turn was based upon two reports appearing in the *Tribun*

Kaltim newspaper on May 5 and 6, 2010.

These sources claimed that the gigantic gecko, supposedly weighing a colossal 64 kg (141 lb), had been captured in a forest by a teenager in Nunukan, just inside Kalimantan (Indonesian Borneo) on the border with the Malaysian state of Sabah on the southeastern Asian island of Borneo. After many people had shown great interest in purchasing it, the gigantic gecko had finally been sold for the eye-watering sum of 64 million Malaysian ringgits (approximately 20 million US dollars) to an Indonesian businessman. He in turn had promptly exited Borneo with his purchase, taking it instead to Kuala Lumpur on the Malaysian mainland.

Since then, nothing more has been heard about this incredible

The (in)famous photo of Indonesia's deceptive giga-gecko as widely present online ("Arbin")

creature—for a very good reason. The entire story was fictitious, and the photograph (claimed in a brief *Jakarta Post* report of May 5, 2010, to have been snapped by someone identified only as "Arbin") was in reality an excellent example of optical trickery. Clearly, the giga-gecko was a hoax, but who had perpetrated it? That remains a mystery.

The gecko specimen in question actually belongs to a very familiar, widely-distributed Asian species known as the tokay gecko, *Gekko gecko*, which is instantly recognizable by virtue of its bluish-grey body liberally patterned with bright red or yellow spots. Although the second largest species of gecko alive today, its maximum total length is a mere 20 inches and its maximum weight no more than 14 ounces. The reason why the specimen in the photograph seems so enormous is that it is sited very much closer to the camera than are the man and the cat sitting on (and under) the railing. This is a classic example of an optical illusion known as forced perspective, often seen in photographs and which, as effectively demonstrated here, can generate some very dramatic (and potentially deceiving) images when purposefully engineered.

Incidentally, while investigating the latter photograph I also discovered online a second, totally separate, but equally striking picture of a normal tokay gecko rendered immense via forced perspective. But in this instance there was no claim that it was a genuine giant specimen. It was contained in a post for August 15, 2012, on the International Gecko and Antique Buyer's website (at: http://geckobuyer.blogspot.co.uk/).

As for the giga-gecko, the final nail in this reptilian riddle's coffin was supplied when a blogger named Abdul Wahid downloaded the gecko photograph directly from the *Tribun Kaltim* newspaper report. For as he revealed in a blog post for May 15, 2010 (at: http://ciucc.wordpress.com/2010/05/15/primacy-effect), he discovered that it was encoded with data revealing that it had been edited with Adobe Photoshop software. Thus, this image not only was an example of forced perspective but also had been photo-manipulated on a computer. Exit the elusive—and illusive—Indonesian giga-gecko!

But just to show that sometimes even the most unlikely, implausible concept turns out to be a reality—in April 2015, British palaeontologist Darren Naish drew my attention (and that of several other zoological colleagues on Facebook) to a remarkable video shot in Indonesia that seemed to show an unfeasibly large yet very much alive and moving tokay gecko being gingerly handled by an unnamed man. The person who had uploaded the video actually owned the lizard and was putting it up for sale. According to a brief statement on his Facebook page, the lizard had supposedly been caught in a forest during a geological survey there. A lengthy ruler laid alongside the lizard showed that it was 138 cm long (i.e. 4.5 feet or 54 inches), i.e. two-and-a-half time longer than the tokay gecko's confirmed maximum total length—indeed, far longer than any species of gecko known to exist today. But that was not the only anomaly concerning this remarkable specimen.

When it moved, its locomotion was much more like that of a monitor lizard than a gecko. Also, it didn't exhibit the geckos' characteristic toe-curling action per step taken; and when the man lifted it up off the floor, there was no resistance, whereas geckos normally grip firmly to any firm substratum via the suckers on their toes. Something was amiss here, but an even bigger surprise was to come.

Investigating tokay gecko videos on YouTube, I found several that featured monitors that had been disguised as tokay geckos (with varying degrees of success), via paint jobs on their bodies, false flanges attached to their toes to make them look more like varanid toes, and, most bizarre of all, false gecko heads placed over their heads. In the video brought to my attention by Darren, the lizard shook its head violently at the beginning, which puzzled all of us at first, but we now realize that what it was doing was attempting to rid itself of the false head placed over its own. Moreover, during the course of the video, the lizard seemed very unsure of where it was going, bumping into various objects, which can now be readily explained as an inevitable result of the poor animal not being able to see very well (if at all) with the false gecko head placed over its own.

But why on earth are monitors being disguised as giant tokay geckos? The answer is that there is a thriving tokay trade in Indonesia, due to traditional folklore beliefs claiming that these lizards bring all manner of health benefits—and the larger the specimen, the higher the price that its seller can command for it. So if lizards much larger than tokays can somehow be made to look like tokays and sold as such to gullible, unknowledgeable, but wealthy buyers, a very lucrative trade will result. And this is indeed what has happened. The things that people will do for money!

Chapter 4:
THE CAMP FIRCOM CADDY CARCASE—
MONSTER OR MONTAGE?

Broken by great waves,
The wavelets flung it here,
This sea-gliding creature,
This strange creature like a weed,
Covered with salt foam,
Torn from the hillocks of rock.
— Hilda Doolittle, *Hermonax*

Few cryptozoologists will be unaware of the Naden Harbour carcass—an enigmatic serpentine animal corpse measuring 10-12 feet long, sporting what looked like a camel-like head, long neck, pectoral flippers or fins, a very elongate body, and a fringed tail-like section that may have been a pair of hind limbs and/or a bona fide tail. It had been removed from the stomach of a sperm whale by flensing (blubber-removing) workers in a whaling station at Naden Harbour in Canada's Queen Charlotte Islands one day in early July 1937, and had then been placed on a table draped with a white cloth and photographed.

Tragically, the carcass is apparently long-vanished, presumably discarded, but three photographs of it remain and they portray a creature that is sufficiently strange in appearance to have incited considerable controversy ever since as to its possible identity. Almost exactly 20 years ago and based upon the surviving photographic evidence, Ed L. Bousfield, currently a Research Associate at Toronto's Royal Ontario Museum, and Paul H. LeBlond, now retired from the Department of Oceanography at the University of British Columbia in Vancouver, designated the Naden Harbour carcass as the type specimen of the longstanding serpentiform mystery beast informally known as Caddy or Cadborosaurus, the Cadboro Bay sea serpent, frequently reported off the northern Pacific coast of Canada and the USA. In a paper constituting a supplement to the inaugural volume of the scientific journal *Amphipacifica*, published on April 20, 1995, based upon this specimen's morphology as seen in the photos they proposed that Caddy was a living, modern-day species of plesiosaur, and they formally named its species *Cadborosaurus willsi*.

Far less familiar than the Naden Harbour carcass photos, conversely, are two Caddy-linked pictures that were first brought to my notice 20 years ago. To my knowledge, they had never previously received any cryptozoological attention, and even today they remain little-publicized. Consequently, this chapter reviews for the very first time in book form the history and most notable opinions that have been offered to date about the tantalizing object(s) that these pictures depict.

Back in the mid-1990s, I was writing the text to my then-forthcoming book, *The Unexplained: An Illustrated Guide to the World's Natural and Paranormal Mysteries*, and Janet Bord of the Fortean Picture Library was supplying me with illustrations for possible inclusion within it.

Unfortunately, she was not able to supply me with any of the Naden Harbour carcass images as these had not been placed with the FPL and because there was some degree of uncertainty concerning who owned their copyright at that time (they are now in the public domain). So although I did document it in my book, I couldn't illustrate my coverage with one of the pictures of it. Nevertheless, Janet *was* able to find a couple of old picture postcards depicting an alleged Caddy carcass washed up at Camp Fircom in British Columbia, Canada, on October 4, 1936 (less than a year before the Naden Harbour carcass was retrieved), and which I had never seen before. Janet had no details concerning these pictures on file other than the handwritten captions that were already printed upon them, and I was unable to uncover any mention of them in any of the sources of Caddy information available to me. (As for the actual postcards themselves, I assume from their style and the rather primitive quality of their photographs that they were originally on sale in the Camp Fircom area not long after the carcass had originally been discovered there.)

Frustratingly, moreover, the deadlines for writing and submitting to the publishers each section of the book's text meant that by the time that I'd received these interesting images, I'd already written and submitted my full quota of allotted text for my book's Caddy entry, so I couldn't have documented them there anyway. All that I could do, which is precisely what I did do, was include the more detailed of the two images (Picture Postcard #1, hereafter PP1), tagged with the following informative caption: "Postcard depicting an unusual marine carcass, possibly a Caddy, that was found on the beach at Camp Fircom, British Columbia, on 4 October 1936."

In truth, however, the more that I looked at these pictures, especially the close-up view afforded by PP1, the more confused I became about what precisely I was looking at, because they certainly didn't resemble the more traditional supposed sea serpent carcasses that wash up from time to time and invariably prove to be the

The Camp Fircom Caddy carcass, Picture Postcard #1 (public domain)

highly decomposed, distorted remains of sharks, whales, or oarfishes. Indeed, by the time that my book was published in 1996, I considered it likely that they showed nothing more than a collection of sea-divulged debris, which may or may not have been artfully arranged by person(s) unknown to look monstrous in every sense, and thence cash in (possibly literally, via the sale of the picture postcards depicting this deceiving creation?) on the tradition of sea monster sightings in this part of the world. Nevertheless, I was pleased to have been able to include at least one of these puzzling pictures in my book, just in case it elicited any responses from readers supplying additional information or opinions relating to it. And sure enough, this is precisely what happened.

During mid-February 1997, I received a detailed report from a then-university zoology student in Southampton, England, giving his opinion as to what PP1 showed. That student is now palaeontologist Darren Naish, who, like me, has long been interested in cryptozoological subjects in addition to mainstream zoology. Having viewed the photo at length in my book, Darren reported that although there were certain superficial similarities to the Naden Harbour carcass (large skull-like object with an apparent eye socket, long thin elongate body with a pair of anterior lateral projections sited where pectoral fins might be expected to be), he considered it to be a hoax—consisting of a montage of objects that he suspected had been deliberately chosen and arranged to give the impression of a carcass.

The supposed skull, he felt, did not actually possess any definite skull characters, and, tellingly, its eye socket, placed in just the right location to resemble a true eye socket was, in Darren's view, the shell of a mussel. As for the long elongate body, he considered this to be the stem of a large plant, probably kelp, with finger-like projections at its distal or "tail" end resembling the root-like holdfasts that anchor kelp to rocks. In short, a collection of marine/beach detritus deliberately positioned to look like a serpentiform monster carcass, thus echoing my own view regarding this.

Mindful that he hadn't seen Picture Postcard #2 (hereafter PP2), I sent Darren a photocopy of it, which he briefly referred to (and he also included sketches of both pictures) within an expanded, illustrated version of the original report that he had previously sent to me, which was published in the summer 1997 issue of *The Cryptozoology Review*, now defunct. In it, he reaffirmed his opinion that the carcass was a composite of kelp, mussel shell, and beach rocks. Interestingly, although I could see why he thought that the eye socket in PP1's depiction of the skull-like object was a mussel shell, in PP2 it seems to me to be a genuine socket, i.e. a hole, because when this picture is enlarged I am sure that the seawater behind the skull-like object can actually be seen through the socket. That aside, however, I do concur and reaffirm that the Camp Fircom Caddy may be monstrous in form but is merely a montage in nature.

The Camp Fircom Caddy carcass, Picture Postcard #2 (public domain)

Even so, are the main components of it truly botanical rather than zoological in identity?

While corresponding with Darren regarding these two pictures, I was also awaiting a response from Paul LeBlond, to whom I had sent photocopies of the pictures, enquiring his opinion as to what they might portray.

In his letter of reply, dated March 3, 1997, Paul noted that he had seen: "...pictures of a lot of Caddy-like carcasses which have usually turned out to be sharks. Most of them look a lot like the Camp Fircom picture." Of particular interest was his comment:

What makes me think that the Camp Fircom carcass is yet another shark is the uniform roundness of the vertebrae, especially as seen in the upper picture [PP1]. The Neah Bay shark bones looked a lot like that: a "log" made of a series of cylindrical vertebrae, without extensions or projections.

Shark remains are sometimes found washed ashore at Neah Bay and elsewhere along Washington's Pacific coast, and needless to say there are many cases on file (from North America and elsewhere around the world) of such remains being mistaken by eyewitnesses for sea serpent carcasses.

Paul also stated that he had forwarded the photocopied pictures to Ed Bousfield, who very kindly wrote to me on August 27, 1997, with his own comments regarding them:

> I tend to agree with Paul that the Camp Fircom carcase is very probably that of a basking shark. Local beach carcasses that have been attributed to "Caddy"-like animals appear similar to the remains of your photograph. Virtually all such remains, reported (with photographs) during the past 70+ years, have proven to be those of the large pelagic shark species common in surface waters of the North American Pacific coastal marine region.
>
> The only photographs considered by us as reliably that of a "Caddy" carcass, are three fairly good images, taken from three different camera angles by two different photographers, at the Naden Harbour whaling station in 1937, and now deposited in the B.C. Provincial Archives here in Victoria.

Two of the three Naden Harbour carcass photos appear in their book *Cadborosaurus: Survivor From the Deep* (1995).

So might the Camp Fircom Caddy carcass be a highly decomposed shark, or does it at least include some shark-derived components within a heterogeneous array of objects?

For a long time, this enigmatic entity attracted little if any additional attention other than its two pictures featuring in a handful of East European cryptozoological websites but with no attendant comments concerning them. In a guest article regarding the Naden Harbour carcass that appeared in Jay Cooney's *Bizarre Zoology* blog on June 17, 2013, however, Florida-based cryptozoologist Scott Mardis did briefly refer to the Camp Fircom carcass and included PP1. After noting Darren's plant-based opinion regarding its composition and then comparing it to some images of basking shark *Cetorhinus maximus* vertebrae, Scott commented: "I'm not so sure, because it looks very basking sharky to me," and I agree that there is indeed a notable degree of similarity between the supposed carcass's elongate body and a shark's vertebral column.

On February 17, 2015, Darren posted his detailed sketch of PP1 on his Facebook page's timeline and tagged me in his post. He also now opined that the carcass's body certainly resembled a shark's vertebral column (thus updating his original identification of it as a possible plant stem back in his article from 1997), but he remained unsure as to the nature of the carcass's other components. This elicited on my own Facebook page's timeline a number of detailed responses from German cryptozoological researcher Markus Bühler, who illustrated them with relevant images obtained online. Like Paul, Ed, Scott, myself, and now Darren too, Markus favored a shark identity for at least some of the objects constituting the Camp Fircom Caddy carcass, and I am summarizing below the various points that he raised relative to this.

Regarding the carcass's supposed skull, Markus considered that PP1 possibly does show a cranium with a hole, but in a predominantly dorsal view, so that the hole is not an eye socket but is instead the epiphyseal foramen (a large dorsally-sited cranial opening that houses the pineal body in living sharks). If so, then the projections above and below it could be the upper parts of the laterally sited eye orbits. He also noted that the skull might be from a shark but not a basking shark, perhaps instead from a species with different cranial proportions from those of a basking shark, thus explaining why it does not provide an exact match with a basking shark cranium.

Markus considered that shark-derived contributions to the carcass might principally consist of its cranium and vertebral column, but he did also wonder whether, if so, the finger-like projections at the right-hand side of the carcass's body, originally labeled as kelp holdfasts by Darren, may be parts of the shark's fin rays, and he posted some online photos of a fully defleshed shark carcass found underwater that bore exposed fin rays resembling the "fingers" of the Camp Fircom conglomerate. Also, as he correctly pointed out, in some shark species the spinal column between cranium and caudal fin is surprisingly short, so these "fingers" may be exposed rays from the lower lobe of the shark's caudal fin.

Concluding the Camp Fircom carcass discussion thread on my

Facebook timeline, Darren reflected that he'd never considered that a shark might have contributed to this creation when preparing his original article, but he still felt that its overall appearance was the result of an assortment of debris and that this was the key point. That is, the alleged Camp Fircom Caddy carcass was merely a conglomeration of objects from different sources, not a single entity—and I agree entirely with this assessment.

Regardless of whether its body derives from kelp or a shark, or whether its "fingers" are holdfasts or fin rays, or whether its skull is a rock or a shark cranium, or whether the latter object's hole is an eye socket or an epiphyseal foramen or even just a deceptive mussel shell, there can be no doubt that what the Camp Fircom composite is *not*, and never could be, is a deceased Caddy. In short, this is one cryptozoological carcass (and mystery) that, finally, not so much rests in peace as in pieces—very different pieces from a range of very different origins.

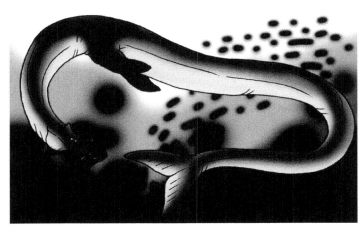

Reconstruction of the possible appearance in life of an adult female Caddy (Tim Morris)

Chapter 5:
ROTHSCHILD'S LOST TUSK AND THE NON-EXISTENT ELEPHANT PIG—AN ENDURING CRYPTOZOOLOGICAL CONUNDRUM FROM AFRICA

> *I presented to the Academy of Sciences a note from Mr. Maurice de Rothschild regarding the exploration that he has undertaken in East Africa with Messrs Henry Neuville, Bogek, and Victor Chollet. Among all of the interesting items reported during this exploration, there is a tusk in an excellent state of preservation, which does not resemble any tooth of a fossil or living animal known up to the present. I showed it at the Academy. If it had been found in a fossil state, nobody would have hesitated to consider it as belonging to a new species.*
>
> — Albert Gaudry, in: Maurice de Rothschild and Henri Neuville, "Sur un Dent d'Origine Énigmatique," *Archives de Zoologie Expérimentale et Générale,* Series 4 (October 15, 1907)

I first came upon this very intriguing cryptozoological case many years ago when browsing through Bernard Heuvelmans's book *Les Derniers Dragons d'Afrique* (1978), and it has fascinated me ever since. Recently, I obtained some additional, previously undisclosed information concerning it that turned out to be something of a conundrum in itself. Consequently, it is high time that I finally documented in book form the remarkable story behind one of the most intriguing and potentially significant examples on record of an apparently lost specimen of possible cryptozoological relevance. So here it is—the story, that is, not the specimen, sadly!

It all began in an Ethiopian ivory market, one of many in this country's capital, Addis Ababa, back in the opening years of the 20th century (when Ethiopia was still called Abyssinia). In 1904, Baron Maurice de Rothschild and French zoologist Henri Neuville were visiting this particular ivory market during an East African expedition when they noticed a very odd-looking tusk on a stall owned by some ivory merchants from India. Although in excellent condition, supposedly of modern age (i.e. not fossilized, though possibly several centuries old), and superficially similar to an elephant tusk but with a very dark patina, it only measured 0.56 m (22 inches) long in a straight line and only

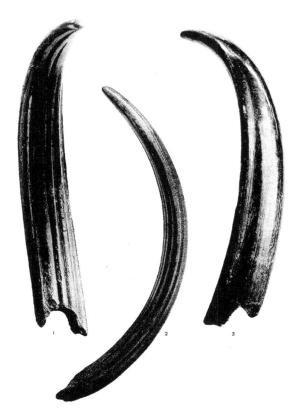

Three different
views of the
mystery tusk
purchased in
Addis Ababa
by Rothschild
and Neuville,
as depicted in
their 1907 paper
documenting it
(public domain)

0.74 m (29 inches) following its curve. So it was smaller than most elephant tusks of comparable proportions that they had previously sold. In addition, it bore a series of longitudinal, regularly spaced, narrow grooves or corrugations on what appeared to be its upper side and a single long but very broad groove on its apparent underside, instead of being smooth-surfaced like regular elephant tusks. Consequently, it had not attracted interest from potential buyers.

The ivory merchants were unable to provide any information regarding this odd tusk's original geographical provenance or from what type of animal it had been obtained. Indeed, they considered it not to be a tusk at all, but rather some form of horn. Nevertheless, intrigued by its strangeness, and no doubt encouraged by its lower-than-normal price, Rothschild and Neuville duly purchased it. Moreover, while still in East Africa, Neuville was informed by some Somali hunters and camel herders that tusks like this one came from an aquatic, hippopotamus-sized creature of great strength, whose tusks curved downwards to the ground, and which inhabited certain very large East African lakes. According to them, one of these lakes is what is now known as Lake Abaja (formerly called Lake Margherita) where they claimed to have seen a living specimen, and another such lake is situated on the border of Kenya and Uganda.

Following their return to Europe, Rothschild and Neuville's unusual zoological purchase was exhibited at three separate scientific meetings. Namely, the Société Philomathique of Paris on January 14, 1905; the Zoological Society of London on November 14, 1905 (where it was presented by Lord Walter Rothschild—distantly related to Maurice de Rothschild—who was a lifelong zoological researcher and prodigious collector of wildlife specimens that he housed at his own personal natural history museum at Tring in Hertfordshire); and the Academy of Sciences, Paris, on December 11, 1905. (Oddly, a short notice in the Zoological Society's *Proceedings* documenting the tusk's exhibition there claimed that two such tusks, not one, had been obtained by Rothschild and Neuville and had been exhibited at the Society.)

Rothschild and Neuville then prepared an extensive scientific paper documenting their anomalous tusk, which was published on October 15, 1907 in the French journal *Archives de Zoologie Expérimentale et Générale* (4th Series), was over 50 pages long, and was aptly entitled "Sur une Dent d'Origine Énigmatique." In it, they compared the tusk closely with a range of others, including those of elephants, hippos, walruses, and wild pigs, as well as a large selection of freak, malformed elephant tusks housed at London's Natural History Museum and elsewhere. However, Rothschild and Neuville claimed that, not only externally but also in terms of its internal grain structure as seen when viewed in cross-section, it differed markedly from all of them, and even from known fossil species of tusked mammal—to such an extent that they deemed it plausible that this unique specimen did indeed derive from some still-undiscovered animal species, exactly as claimed by the Somalis.

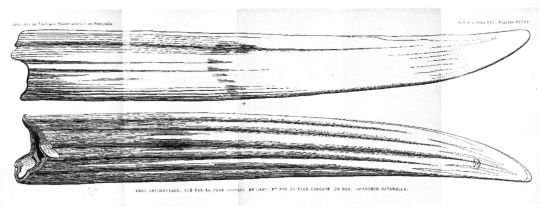

Line illustration of the Rothschild-Neuville mystery tusk, from their 1907 paper documenting it (public domain)

Although referred to for the sake of convenience as a tusk both here and also in the Rothschild-Neuville paper, this enigmatic object was technically a tusk section, not a complete tusk, having been sectioned some way between its tip and its base, with the latter not being present. From their examination of it, Rothschild and Neuville considered that had it been present in its entirety, it would have been almost semi-circular in shape, and that the section preserved was probably derived from a right-hand tusk. Intriguingly, the tip showed no sign of having been sharpened, whereas elephants tend to sharpen their tusks' tips on hard objects that they encounter. From the trace of the pulp cavity remaining, the authors considered it likely that their perplexing tusk had exhibited continuous growth in life, like elephant tusks, rather than like normal teeth that do not grow continuously through life.

The Rothschild-Neuville mystery tusk was totally devoid of enamel (moreover, it occurs only at the onset of tusk development in elephants, whereas a thick layer is present in hippo tusks; enamel occurs at the tip in walrus tusks, though this tends to be worn off as the animal matures). However, there was a layer of cement, which, although not particularly thick, was very extensive (a cement layer replaces the enamel layer in elephant tusk development). As for its anomalous grooves, superficially similar structures are also borne upon the surface of hippo tusks, but their arrangement is very different from the grooves of the Rothschild-

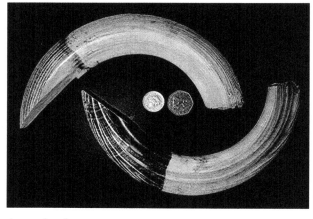

Neuville mystery tusk. This is also true of the grooves that sometimes occur on wild pig tusks, and those present on various teratological tusks from elephants that were examined by the authors. Several distinctive traits can be found in diseased examples of elephant tusks, but their mystery tusk exhibited none of those traits; in fact, it appeared to be perfectly healthy, thus eliminating another possible explanation for it. (For further morphological and

A couple of hippopotamus tusks, the upper one revealing the grooves on its upper side, the lower one revealing those on its underside (Karl Shuker)

histological details, please consult the meticulous description in their original paper.)

Having assessed all of their mystery tusk's characteristics, Rothschild and Neuville concluded that it compared more closely with elephant tusks than with those of any other mammal, but that even with these it also exhibited notable differences. Consequently, their preferred hypothesis was that it had therefore originated from a species of large mammal still unknown to science but probably most closely related to proboscideans (Proboscidea is the taxonomic order containing elephants).

Notwithstanding this exciting prospect, the Rothschild-Neuville mystery tusk rapidly sank into zoological obscurity following their paper's publication in October 1907, with only Heuvelmans's aforementioned documentation of it in his 1978 book briefly raising its profile. And yet with modern-day advances in microscopical, chemical, and genetic analyses readily available, surely a re-examination of this specimen utilizing these techniques might unlock the secret of its original owner's still-cryptic taxonomic identity? Quite possibly, if only this mystery tusk's current whereabouts were known.

French cryptozoologist Michel Raynal was in direct, frequent correspondence with Heuvelmans for many years, and I recently learnt from him that Rothschild and Neuville had deposited their mystery tusk at France's National Museum of Natural History in Paris and that it had been accorded its own accession number. However, he had been told by Heuvelmans about 20 years ago (i.e. 7-8 years before Heuvelmans died in 2001) that the museum staff could not find the tusk when he (Heuvelmans) had requested sight of it, not even when he had given them its correct accession number. So it had probably been lost many years earlier. But what was Heuvelmans's opinion concerning the possible identity of the mysterious creature from which the Rothschild-Neuville tusk had originated? This is where the story becomes even more intriguing, because Heuvelmans expressed two very different opinions concerning this issue—one publicly, another one privately.

The more conservative (relatively speaking) of the two is the opinion that he expressed in his book *Les Derniers Dragons d'Afrique* (1978). Here he suggested that the unrevealed tusk-bearer might be an unknown species of proboscidean. In his classic book *On the Track of Unknown Animals* (1958), Heuvelmans had referred to a cryptic aquatic proboscidean, the so-called water elephant (aka river elephant), which I have also reviewed in various of my own writings, particularly in my book *In Search of Prehistoric Survivors* (1995) and on my *ShukerNature* blog.

Said to inhabit certain Congolese lakes (Lord Walter Rothschild had collected reports appertaining to this cryptid there and also in Tanzania as well as South Africa, as noted in the Rothschild-Neuville paper), and sighted on two separate occasions during the first decade of the 20th century by French explorer M. Le Petit, the water elephant, as its name suggests, allegedly spends much of its time submerged, only emerging onto land during the evening. It is readily distinguished from normal elephants morphologically too, via its very elongate, ovoid head and its extremely short, tapir-like trunk. However, it is also said to lack tusks. If so, while again distinguishing it from normal elephants, by definition this characteristic also eliminates it from consideration as the originator of the Rothschild-Neuville mystery tusk.

Nevertheless, following publication of the Rothschild-Neuville paper, Le Petit was urged by Paris's National Museum of Natural History to pursue reports of a similar creature supposedly inhabiting Lake Chad in West Africa. Moreover, according to an article penned by Emile Trouessart in the January 21, 1911 issue of the French periodical *La Nature*, Le Petit would be accompanied

Portrayal of the Congolese water elephant (Markus Bühler)

in the search by noted French zoologist Emile Gromier; it also suggested that both the West African water elephant and the unknown originator of the Rothschild-Neuville mystery tusk might be a living deinothere.

Deinotheres were very large, highly unusual proboscideans known only from fossils, and characterized by their lower jaw's downward-curving pair of tusks. In their paper, Rothschild and Neuville had discounted a deinothere as a plausible identity for their tusk's originator, but Heuvelmans did not dismiss a relatively small, amphibious, scientifically undetected modern-day version out of hand when documenting the water elephant and the Rothschild-Neuville tusk in his books. As for the planned expedition to Lake Chad: I have been unable to discover whether Le Petit and Gromier ever did go there—but even if they did, they certainly didn't find a living deinothere.

Heuvelmans's second, privately expressed opinion as to what creature the Rothschild-Neuville mystery tusk might have come from is even more radical than an undiscovered species of probiscidean. In a couple of emails sent to me by Michel Raynal during June 2014, Michel stated:

> In Heuvelmans's dossier on the tooth in Lausanne [the site of his Centre de Cryptozoologie], I noticed that there is an article on

Afrochoerus, making me think that Heuvelmans believed that the "owner" of this tooth was more or less related to this fossil suid...

Heuvelmans seemed to believe that it was a living *Afrochoerus* (he even told me, when I visited his "centre de cryptozoologie" in the 1980s, "je sais ce que c'est" = I know what it is).

However, *Afrochoerus* tusks are somewhat different.

They are indeed! In fact, the tusks of this African fossil pig once incited a major controversy of their own.

Dating from the Pliocene epoch, *Afrochoerus nicoli* was formally described in 1942 by celebrated palaeontologist Louis Leakey from the famous Olduvai fossil deposits in Tanzania, with substantial additional finds excavated there in subsequent years. Its most striking features were the tusks originally uncovered with it, which were disproportionately long. In early reconstructions of this species, which portrayed it with these huge tusks pointing forward, it resembled a bizarre hybrid of wild pig and mastodont!

In fact, this was quite prophetic, as palaeontologists now know that these tusks weren't suid (pig) tusks at all, but were in reality from *Stegotetrabelodon*—a genus of fossil gomphothere elephant! What had happened was that *Afrochoerus* had been based upon cranial remains of a bona fide fossil suid called *Metridiochoerus compactus*, the giant warthog, that had been discovered in close proximity to some tusks from *Stegotetrabelodon*, but which had been mistakenly assumed to be all from the same animal. In short, *Afrochoerus*—the extraordinary "elephant pig"—was a non-existent composite, whereas the real tusks of *M. compactus* are nothing more than slightly larger versions of those possessed by modern-day warthogs, and project sideways, not forwards.

Consequently, although he apparently did not realize it, by favoring *Afrochoerus* as the identity of the Rothschild-Neuville tusk, Heuvelmans was still supporting a proboscidean candidate (albeit a

Afrochoerus purposefully portrayed here as originally conceived when first discovered, i.e. erroneously sporting the tusks of a prehistoric elephant (Hodari Nundu)

Accurate reconstruction of *Metridiochoerus* (Nobu Tamura/ Wikipedia)

different one from the deinothere option), because the tusks to which he was comparing the latter mystery tusk were actually elephant tusks, not pig tusks.

But could this mystery tusk have derived from something other than a proboscidean? When I inspected the illustrations of it in the Rothschild-Neuville paper, images of walrus tusks immediately came to mind, even though the authors had discounted this possibility in their paper. True, there are certain discrepancies. The underside of walrus tusks have a single wide groove (just like the mystery tusk), but the upper side typically bears only a series of very fine longitudinal cracks of varying lengths. Also, walrus tusks are generally long and fairly straight, whereas if Rothschild and Neuville were correct in assuming that in its entirely their tusk would have been almost semi-circular in shape, this would also argue against a walrus tusk as its identity. So too does the dark brown patina exhibited by their tusk.

However, I have seen photos of walrus tusks that do bear deep longitudinal grooves. Moreover, when I consulted Paolo Viscardi, the natural history curator at London's Horniman Museum, who also favors a walrus tusk as the best candidate for this enigmatic specimen, he stated:

> Walrus tusks can also be quite strongly curved and if they [Rothschild and Neuville] were extrapolating the curve and complete length based on pulp cavity extent beyond the sawn region, they may have been wildly out if they didn't have Walrus in mind, since Walrus cavity space is less extensive than what we see in something like an Elephant or Hippo.

And although the authors stated that it was not a fossil tusk, fossil walrus tusks do sometimes have a dark patina.

But how would a walrus tusk have reached the ivory market of Addis Ababa anyway, bearing in mind that walruses only occur in the Arctic Ocean and the subarctic zones of the Atlantic and Pacific Oceans? Fellow Rothschild-Neuville tusk researcher Matt Salusbury, author of the book *Pygmy Elephants* (2013) in which this specimen is briefly mentioned, suggested to me in an email of June 30, 2014 that whalers and immigrants from Scandinavia en route to South Africa just prior to the tusk's appearance in Addis Ababa might have brought it with them.

Intriguingly, in South Africa's Orange Free State there are ancient bushman cave paintings that portray an unidentified creature bearing a remarkable resemblance to a walrus, and which has been colloquially dubbed the jungle walrus. The most famous one (portrayed here), which appeared in *Rock-Paintings in South Africa* (1930) by George W. Stow and Dorothea F. Bleek, can be found in a cave in Brakfontein Ridge, which was at one time (and perhaps still is) contained within the grounds of a farm called "La Belle France." Just an imaginary beast, or a bona fide cryptid with possible relevance to the Rothschild-Neuville tusk?

Of course, as noted earlier previously, if only the Rothschild-Neuville tusk could be scientifically examined utilizing the plethora of modern-day techniques readily available nowadays, its secrets would soon be teased out and laid bare. Yet as also noted previously, it was apparently lost long ago somewhere within the labyrinthine interior of its final resting place, the National Museum of Natural History in Paris—or was it?

Matt Salusbury has informed me that he is currently pursuing the tantalizing possibility that the tusk has not been mislaid but simply

The "jungle walrus" depicted in a cave painting at Brakfontein Ridge, South Africa (public domain)

mislabeled. He is hoping to have its accession number rechecked and also those of comparable specimens there, such as walrus tusks, to determine whether this most mystifying specimen is actually hiding in plain sight—not lost at all, therefore, merely overlooked. He is also checking the accession logs at London's Natural History Museum, the Zoological Society of London, and the Tring Natural History Museum, just in case the tusk was transferred from Paris to any of these English repositories.

I wish Matt every success in his exciting undertaking, and very much hope to be able to announce soon that the Rothschild-Neuville mystery tusk has indeed resurfaced—which would be a major success for contemporary cryptozoology.

Chapter 6:
MYSTERY FROGS AND TOADS —SEEKING LOST AND UNIDENTIFIED SPECIES

And Garp and Helen and Duncan held their breath; they realized that all these years Walt had been dreading a giant toad, lurking offshore, waiting to suck him under and drag him out to sea. The terrible Under Toad.
Garp tried to imagine it with him. Would it ever surface? Did it ever float? Or was it always down under, slimy and bloated and ever-watchful for ankles its coated tongue could snare? The vile Under Toad.
— John Irving, *The World According To Garp*

New species of frog and toad are still being discovered every year, and in modern times some of them have proved to be very dramatic morphologically and/or behaviorally, such as the "living fossil" purple frog *Nasikabatrachus sahyadrensis* of India's Western Ghats, the gastric-brooding platypus frogs *Rheobatrachus silus* and *R. vitellinus* of Australia, Cambodia's green-blooded turquoise-boned Samkos bush frog *Chiromantis samkosensis*, and the vampire-fanged flying frog *Rhacophorus vampyrus* of Vietnam. Moreover, browsing through the chronicles of cryptozoology yields a wide selection of cases suggesting that several more decidedly noteworthy species might still await discovery out there, as well as a number of famously lost species in sore need of rediscovery.

GIANTS AROUND THE GLOBE

The 20th century opened with one of the most dramatic amphibian discoveries of all time—the aptly named goliath frog *Conraua goliath*. Up to 14.5 inches long from snout to vent, plus a pair of enormous hind legs, and weighing up to 8 pounds, it is the world's biggest known species of living frog (or toad, for that matter), as large in fact as certain antelopes with which it shares its Middle African domain. However, another very sizeable African frog remains unidentified and unexamined by science almost 70 years after it was first documented.

On December 31, 1945, an article penned by Harvard University herpetologist Arthur Loveridge was published in the zoological journal *Copeia*, concerning an attack some months earlier upon an askari (native policeman) at Tapili, Niangara, in what was then the Belgian Congo (later renamed Zaire, and now called the Democratic Republic of Congo). Loveridge's source of information concerning this incident was a Mr. C. Caseleyr, then Administrator of the Niangara Territory. The askari had come to Caseleyr to inform him that while walking by a pool earlier that evening, he had been bitten on the leg by what proved to be a very large frog—he'd lunged out at his attacker with a large club that he was carrying and had killed it outright. And as conclusive proof of his statement, the askari had brought with him the frog's body to show it to Caseleyr.

What was even more surprising than the fact that he'd been attacked by such a creature, however, was the wound that it had produced. For when Caseleyr examined the askari's leg, he could see two puncture marks resembling the wounds that a dog's teeth would leave—and that was not all. When he then examined the dead frog, Caseleyr was very startled to discover that its upper jaw and lower jaw each possessed a pair of sharp teeth that actually did resemble a dog's canine teeth, and its tongue was forked like a snake's. The frog was grey-green in color dorsally, with a large orange patch on its chest and stomach, and in general shape and size was very large, broad, and fat (though no specific measurements for it were provided by Loveridge).

Loveridge referred to this animal throughout his article as a frog, but Caseleyr personally felt that it seemed more like a toad, although he freely confessed that he had no informed knowledge on such matters. However, Loveridge stated that the creature's notched tongue eliminated a toad as its identity. Yet he also recognized that its description (and most especially the presence of two teeth-like projections in its upper jaw, not just in its lower jaw) readily differentiated this sizeable mystery frog from all known species in that region. Consequently, albeit with great hesitation, Loveridge concluded his article by speculating that "it may be that some large species remains to be discovered in the Niangara region." Yet if so, it has somehow managed to elude scientific detection for several decades since then (and continues to do so today). Impossible? Perhaps not...

A team of nine scientists from Peking (now Beijing) University, led by 58-year-old biologist Chen Mok Chun, travelled to some very large, deep mountain pools (one of which is named Bao Fung Lake) in Wuhnan within central China's Hubei Province during August 1987 in order to film the area and its wildlife. While setting up their television cameras, however, they were allegedly treated to an exhibition of local wildlife far beyond anything that even their wildest imaginations—or worst nightmares—could have conceived.

In full view of the scientists, three huge creatures supposedly rose up out of one of these pools and moved towards the pool's edge nearest to them. The stunned eyewitnesses likened these grotesque monsters to giant toads, with grayish-white skins, mouths that were said to be about 6 feet across, and eyes "bigger than rice bowls."

According to Chen, they silently watched the scientists for a short time.

The author alongside a statue of a hypothetical giant frog (Karl Shuker)

Then one of them opened its huge mouth and swiftly extended an enormously long tongue, which it wrapped around the cameras on tripods. As it promptly engulfed the tripods, its two companions let forth some eldritch screams, and then all three monsters submerged, disappearing from view.

The delayed-shock reaction experienced by the scientists was so great that one of them dropped to his knees and was physically sick, according to Chinese reports, which were summarized in various overseas media accounts. This incident seems so utterly incredible that one would surely feel justified in dismissing it entirely as a hoax were it not for the fact that the eyewitnesses in question were all trained scientists, including a major name in Chinese biological research, and all from the country's leading university. Moreover, reports of such creatures being sighted in this same locality by local fishermen date back at least as far as 1962.

The native Indians inhabiting various tropical valleys in the Chilean and Peruvian Andes frequently report the existence of a greatly feared, giant form of toad called the sapo de loma ("toad of the hills"). It is said to be deadly poisonous and capable of preying upon creatures as large as medium-sized birds and rodents.

Science has yet to examine a sapo de loma, but as I have pointed out in my book *The Encyclopaedia of New and Rediscovered Animals* (2012)), there is a major precedent for discovering batrachian behemoths in

South America. Rolf Blomberg (1912-1996) was a Swedish explorer, photographer, and writer, and in 1950 he was instrumental in bringing to scientific attention a hitherto-undescribed species of giant toad native to southwestern Colombia. He achieved this significant feat by capturing a huge specimen there that he brought back with him to his home in neighboring Ecuador, and which became the type specimen of this spectacular new species—duly dubbed *Bufo blombergi* in his honor a year later. Blomberg's giant toad (aka the Colombian giant toad) can attain a total snout-to-vent length of up to 10 inches, which is greater even than that of the heavier, more massively built cane toad *B. marinus* (=*Rhinella marina*), the world's largest toad species.

Incidentally, German cryptozoological correspondent Markus Bühler recently informed me that while browsing through a series of yearbooks from the 1970s for Stuttgart's Wilhelma Zoo, he noticed that the 1971 volume mentioned the chance birth at the zoo some years earlier of hybrids between a male *B. marinus* and a female *B. blombergi*. They were apparently indistinguishable from the paternal species, but exhibited unusually strong growth—no doubt a result of hybrid vigor. Bearing in mind that they were crossbreeds of the world's largest toad species and the world's longest toad species, it is a great pity that the book did not contain any additional information concerning them, because they must surely have had the potential to become veritable mega-toads!

In any event, mindful of the relatively belated scientific discovery of Blomberg's giant toad it would be rash to rule out entirely the possibility that a comparable species still awaits discovery in remote, rarely-visited Andean valleys not too far to the south of the latter species' distribution range.

Another noteworthy precedent for such finds occurred during the early 1960s in New Guinea, although it did not attract widespread attention outside herpetological circles until I documented it in my book *The Lost Ark* (1993). In December 1960, news emerged concerning the discovery of what was said to be an exceptionally large form of frog here. Known locally as the agak or carn-pnag, it supposedly measured up to 15 inches from snout to vent, thereby rivaling Africa's goliath frog. Three years later, however, this would-be usurper of the latter species' long-held record was exposed as a charlatan, when it was officially described by Australian zoologist Michael Tyler. Naming it *Rana jimiensis* (later renamed *Hylarana jimiensis* by some workers), Tyler revealed that its precise provenance was Manjim, on the Jimi River Valley's Ganza River in Papua New Guinea's Western Highlands province, and that the first reports concerning its size had been exaggerated. The maximum authenticated length on record for the agak constituted a rather more modest 6.5 inches. Even so, this still makes it the second largest species of frog native to New Guinea— exceeded only by the Arfak Mountains frog *R. arfaki*, whose females can attain a snout-vent length of eight inches.

Meanwhile, back in South America there are also claims among the Mapuche people relating to a supposedly immense species of toad indigenous to various lakes, lagoons, and irrigation channels in southern Chile and southwestern Argentina. Known as the arumco ("big water frog") in Argentina and the vilú in Chile, it is said to measure up to three feet long. One such creature, dwelling in a lake, reputedly devoured a horse that was attempting to wade across, but whether this story is true or merely native folklore remains unresolved.

THE FROG WITH THE LUMINOUS NOSE

In one of his famous nonsense poems, Edward Lear wrote about a fictitious creature known as the dong with the luminous nose, and our next mystery beast was apparently a frog with a similar attribute, yet it was anything but nonsensical or fictitious.

While visiting an animal fair at Newton Abbot in Devon, England, during June 1997, Devon-based cryptozoologist Jonathan Downes noticed a cage containing some tree frogs, reputedly from northern Cameroon in Middle Africa and priced at £25 per frog. Jonathan was intrigued by these creatures, which each bore a blue spot on its snout, because he could not identify their species. He was even more intrigued when informed by their vendor that the blue spot on their snout glowed in the dark like a flashlight, possibly to attract insects as prey.

Artistic representation of the "flashlight tree frog" (Connor Lachmanec)

Nevertheless, Jonathan felt that £25 was too expensive a sum, so he resisted the temptation to buy one—a decision that he would soon bitterly regret. For when he later described these curious "flashlight tree frogs" to various herpetological colleagues, he was shocked to discover that there is no species of frog known to science that can glow. Consequently, he had missed the opportunity to purchase a specimen of what might not only be a completely unknown species, but also exhibit a talent unique even among the world's considerably varied array of frogs.

A PHANTOM WHITE FROG AND A BLIND WHITE FROG

The year 1997 saw the formal description of a new species of frog characterized by its skin's distinctive spectral sheen. Its holotype was a female specimen discovered several years earlier on the rocky banks of a forest stream in the mountain forests of southern Costa Rica by Karen R. Lips from Saint Lawrence University on New York and Jay M. Savage from Miami University in Florida. This very memorable species was fittingly christened *Eleutherodactylus phasma*, the phantom frog, due to the ghostly pallor of its almost pure white skin, but it was later reassigned to the genus *Craugastor*, becoming *Craugastor phasma*. Nearly two inches long, it was initially thought to be a freak albino specimen of a previously described *Eleutherodactylus* species, but this was later disproved. Sadly, however, almost 20 years later this mystifying species is still known only from its holotype because no others have ever been sighted.

Speaking of mystifying white frogs, I also have on file a tantalizing snippet of text mentioning what may be a very significant but still-undiscovered species in Rwanda, eastern Africa. It featured in *Among Pygmies and Gorillas* (1923), a book by Prince William of Sweden, which detailed his explorations of Rwanda as part of the Swedish Zoological Expedition to Central Africa that he led in 1921. The passage in question reads as follows:

> The white, or blind, frog which was said to live in Mutanda was also conspicuous by its absence. We hired the natives to collect amphibians from the whole of the lake, and they came regularly once a day with big baskets absolutely full of frogs. In hundreds and thousands, in cubic yards. The whole camp was alive with frogs. In vain! There were no white ones among them, and every one had two big staring eyes.

Could this mysterious blind white frog have been a specialized cave-dwelling species? If so, it would compare with various spelaean salamanders and freshwater fishes that also lack eyes and skin pigment, as these are superfluous when living in a permanently lightless world.

TIME TO SNIFF OUT THE SKUNK FROG?

The phantom frog of Costa Rica is by no means the only notable species of lost frog in dire need of rediscovery. Indeed, there are now a great many (including both species of the remarkable gastric-brooding or platypus frog, tragically), which is due in no small way to the devastating effects of the potentially lethal skin disease chytridiomycosis. Spread by the fungus *Batrachochytrium dendrobatidis*, it has decimated amphibians worldwide (having been detected on at least 287 different species in 36 different countries) since it was first identified in 1998. One of its most spectacular victims was the Panamanian golden frog *Atelopus zeteki* (actually a toad!), which became extinct in the wild shortly after this deadly disease was first recorded in its valley habitat in 2006, but it is being successfully bred in captivity. Another one, equally eye-catching, was Costa Rica's golden toad of Monteverde *Incilius periglenes*, which went from being abundant to extinct within the space of just two years during the mid-1980s. Tragically, there are no individuals of this beautiful species in captivity.

There is even an entire book devoted to these missing creatures. Written by Robin Moore and published in 2014, it is entitled *In Search of Lost Frog*s, and includes a number of exotic, elusive species. Happily, a few have been successfully rediscovered in recent years, such as Costa Rica's variable harlequin toad *Atelopus varius* (in 2003), and Israel's Hula painted frog *Latonia nigriventer* (in 2011). Various others, conversely, like Colombia's beaked toad *Rhinella rostrata* (still known only from two specimens collected back in 1914), remain tenaciously awol despite being specifically searched for.

One of the most enigmatic and fascinating of all lost frogs must surely be South America's very aptly named skunk frog. This yellow-patterned, green-skinned species of arrow-poison frog was discovered in 1981 by Venezuelan biologist Alfredo Paolillo, within the Andean cloud forests of the extreme northeastern corner of Trujillo, northwestern Venezuela. In stark contrast to other arrow-poison frogs, this particular species was nocturnal rather than diurnal, and was principally aquatic (with fully webbed feet) rather than terrestrial. Moreover, measuring up to 2.4 inches long it was unexpectedly large for an arrow-poison frog (indeed, it is by far the largest species currently known to science).

The skunk frog's most distinctive characteristic, however, which has earned it its English name, was only revealed when scientists attempted to handle it—swiftly registering its disapproval, this extraordinary creature secreted a liquid whose smell was more than just foul; it actually resembled the infamous stench of the anal liquid ejected by threatened skunks.

So far removed taxonomically from other arrow-poison frogs that it was housed in a new genus and a new taxonomic family, the skunk frog was scientifically described in March 1991, and as a reminder of two of its most characteristic features it was named *Aromobates nocturnus*. Worryingly, however, this remarkable species has not been seen since its discovery, despite being searched for, and is classed as critically endangered by the International Union for Conservation of Nature (IUCN) and may even be extinct.

WINGED TOAD OR FLYING GURNARD?

In my book *The Menagerie of Marvels* (2014), I documented some extraordinary reports of mystifying, still-unidentified creatures claimed to be bona fide winged toads. Since then, one additional report has been brought to my attention by British cryptozoological researcher Richard Muirhead, but in this instance I am confident that the creature in question can be readily identified with a known animal form.

The report, which sadly lacked any pictures of the animal that it described, had appeared on August 28, 1869, in an English newspaper called the *Grantham Journal*:

> An American paper reports the capture of a flying toad at Cape Henry [on Virginia's Atlantic shore] a few days ago, and says it is now in Washington. It is of most singular conformation and of beautiful variated hues, measuring about six inches in length, with a perfectly flat, bony back, eyes wide spaced and in the centre of a...mouth, and fins as large as wings about the centre of the body on each side.

An account had also appeared in the *Birmingham Gazette* on July 28, 1869. The reference to this specimen being "now in Washington" suggests that following its capture it was sent to the Smithsonian Institution, so that would seem to be the most promising place to contact in the hope of uncovering further information concerning this strange creature. Having said that, however, I feel sure that we can ascertain its identity with a high degree of probability just from the report alone. For although the above account may sound truly bizarre, it is in fact an accurate description of an extremely distinctive and quite spectacular species of small marine fish—*Dactylopterus volitans*, the so-called flying gurnard.

Flying gurnard, as portrayed in Marcus Bloch's 12-volume *Encyclopedia of Fishes*, 1782-1795 (public domain)

Flying gurnards, of which there are several species, are exceptionally difficult to classify. Measuring up to one foot long, the Atlantic flying gurnard *Dactylopterus volitans* (whose distribution along the USA's eastern coast includes Virginia, where the alleged flying toad was captured) and the comparably sized starry flying gurnard *Dactyloptena* (=*Daicocus*) *peterseni* from the Indo-Pacific are among the most familiar representatives of these curious fishes. Over the course of time, they have been variously categorized with the true gurnards, the sea-robins, and the sea-horses, but are nowadays generally housed in their own suborder within the taxonomic order Scorpaeniformes, or even within an order entirely to themselves. Apart from *Dactylopterus volitans*, flying gurnards are mostly of Indo-Pacific distribution, and are benthic (bottom-dwelling) species.

Superficially similar to true gurnards, but distinguished from them anatomically by subtle differences in the arrangement of their head bones and the spines of their pectoral fins, flying gurnards are characterized by their large, bulky heads, encased in hard bone and surprisingly toad-like in appearance when viewed both in profile and face-on; their brightly-colored, box-shaped bodies, dappled with multi-hued spots; and, above all else, by the enormously enlarged pectoral fins of the adults, expanded like giant, heavily-ribbed fans or paired wings. Compare this description with that of the "flying toad" of Cape Henry and it seems evident that they are referring to one and the same creature.

This chapter's search for lost and unidentified frogs and toads has scoured the globe, from Africa and Asia to South America and New Guinea, and even the coastal sea off Virginia, but perhaps some of the most remarkable discoveries might yet be made much closer to home. Over the years, the pet trade has proved a surprisingly successful source of mystifying herpetological specimens that when formally examined by scientists have been found to represent entirely new, undescribed species. Judging from the tantalizing episode of the "flashlight frog" with the glowing nose, it looks possible that future discoveries of this nature will include some very notable new frogs and toads.

And indeed, if any herpetological collectors reading this chapter have seen (or actually purchased!) specimens of the "flashlight tree frog" at pet fairs visited by them lately, I'd be very interested to receive details.

Chapter 7:
THE ZOO THAT NEVER WAS—A MIRAGE OF CRYPTOZOOLOGICAL DELUSIONS AND ILLUSIONS

The pursuit of a mystery is one of those occupations that appear to have a never-failing attraction for the enquiring minds of Western peoples. Advertise sufficiently some inexplicable phenomenon and you will never fail to draw a crowd of sightseers or a rush of letters to the correspondence columns of the daily newspapers. The circumstances of the mystery would appear to matter little beside the fact of the mystery itself and even the greatest minds have been interested in happenings that in themselves were quite ordinary and in their surroundings often squalid and sordid.

— F.B. MacRae, "More African Mysteries,"
National Review (1938)

Tricks you need to transform something which appears fantastic, unbelievable into something plausible, credible, those I learned from journalism. The key is to tell it straight. It is done by reporters and by country folk.

— Gabriel García Márquez interviewed by
Marlise Simons, *New York Times*
(December 5, 1982)

All of the cases documented in this chapter were eventually revealed to be very different indeed from what they initially seemed—much to the chagrin and shame of those who fell foul of their beastly blandishments.

THE STRIPED ANTEATER THAT MADE A BUFFOON OUT OF BUFFON

There are four recognized species of modern-day South American anteater—or vermilinguan, to be taxonomically precise (the unrelated aardvark, the pangolins, and the echidnas are all sometimes referred to colloquially as anteaters too—respectively, the African anteater, the scaly anteaters, and the spiny anteaters). These are: the giant anteater *Myrmecophaga tridactyla*; the pygmy or silky anteater *Cyclopes didactylus*; and somewhat midway in size between these two species, the northern tamandua *Tamandua mexicana* and the southern tamandua *T. tetradactyla* (these two were previously lumped together as a single species, the tamandua *T. tetradactyla*).

Each of the two tamandua species is split into four subspecies, and although the most familiar appearance in both species is one in which the animal possesses a black vest-like coat pattern over its torso, with the remainder of its body and its head of paler coloration, there is much variation in both coat color and pattern.

Variation notwithstanding, however, no tamandua had ever been reported before (or, indeed, has been since) that even remotely resembled a certain extraordinary specimen sent during the 1700s to the pre-eminent French naturalist Georges-Louis Leclerc, Comte de Buffon (1707-1788), for examination. What made it unique was that, totally eschewing the traditional "black vest vs. paler elsewhere"

The fraudulent striped tamandua from Buffon's *Histoire Naturelle*, 1749-1788 (public domain)

tamandua image noted above, this particular individual was very distinctively patterned all over its body, legs, tail, and even its long snout with bold, highly contrasting black and gold stripes!

Needless to say, the Comte de Buffon was captivated by this veritable bumblebee in anteater form, and in 1763 he duly incorporated it as a major new species, the striped tamandua, in his monumental, 36-volume magnum opus, *Histoire Naturelle* (1749-1788). (NB: one of the southern tamandua's four subspecies is also referred to sometimes as the striped tamandua, but it bears no resemblance to the singular specimen documented here, so it should not be confused with this latter animal.) He even commissioned a full-color plate for his *Histoire Naturelle*, portraying his striped tamandua there in all its banded beauty, which to my knowledge is the only depiction of this creature ever produced.

Tragically for the Comte de Buffon's reputation, however, when the striped tamandua's holotype was examined by other zoologists after his death, it was discovered that he had been the victim of a cunning hoax. The creature was not a tamandua at all, but was instead a coati—a long-nosed relative of the raccoons, occurring in three recognized species—

which had been deftly modified to resemble an anteater (even its teeth had been removed), and whose stripes were equally artificial. The perpetrator of this cruel practical joke was never identified, but once their hoax had been exposed, the Comte de Buffon's now-fraudulent striped tamandua made a swift, un-mourned exit from the natural history tomes, never to return.

A BABY CHUPACABRA? YOU'RE HAVING A LAUGH—OR A GAFF!

The chupacabra is unquestionably one of the most famous and iconic mystery beasts of modern times. So when a photograph purportedly depicting a mummified baby chupacabra appeared online, it was certainly going to attract plenty of attention.

Perusing a number of websites containing and discussing it after first hearing about it a few years earlier, I soon learnt that the most commonly repeated claim made was that this supposed chupacabra infant or mini-chupacabra had been discovered under an abandoned barn in a small village in Chiapas, Mexico, during the barn's demolition in or around July 2007, and that it couldn't be identified with any known species of animal because a series of DNA, fur, and tissue tests had all proved inconclusive. The reality, of course, proved to be a totally different matter. The photograph itself (details as to where it can be viewed online will be given a little later in this section) was copyrighted to the oddly named Zeehc H. Ted (but here is an artistic interpretation of the creature in life, as also seen on this present book's front cover).

Representation of the Mexican mini-chupacabra in life (Michael J. Smith)

Even a brief examination of the well-preserved specimen in the photo was enough for me to feel confident that it was a gaff, i.e. a fake taxiderm specimen composed of body parts from various different species deftly combined together. The principal component appeared to be a baby mammalian carnivore, almost certainly a raccoon *Procyon lotor*, judging from its dentition, shape of its paws, and general body and facial conformation. The spines inserted upon its head resembled claws.

In a search of clues to its real origin, I scoured the internet and soon uncovered the truth behind the travesty (albeit an extremely skillfully executed travesty!). Just as I suspected, the "baby chupacabra" was indeed a fake—one of many spectacular examples produced by a wonderfully ingenious gaff creator hailing from the USA and exhibiting photos of his gaffs on the deviantart.com website. There, he memorably refers to himself as "Creator Of Things That Should Not Be," and his user name is dethcheez. This in itself is (or should have been) a major clue to anyone attempting to track down the photo's origin—because the name of the latter's copyright owner, Zeehc H. Ted, is of course dethcheez spelt backwards!

As for the specimen itself, the *DeviantArt* page containing the original photograph of it, which can be directly accessed at http:// dethcheez.deviantart.com/gallery/?offset=96#/dw76ja (the hand holding the specimen was added later to this photo), was uploaded on April 26, 2007 by its creator, dethcheez, who labels it as a "Mummified Baby El Chupacabra Sideshow Gaff Created from 100% Real Parts." But what are these parts? Reading down the comments below the photograph, all is swiftly revealed, because they were correctly identified on December 10, 2009, by a viewer with the user name inkaholic1089, as verified in a comment posted underneath that of inkaholic1089 by dethcheez two days later. Namely, a baby raccoon with the claws of a common American snapping turtle *Chelydra serpentina* used for its spines.

Another controversial carcass is controversial no longer—and can thus be filed away alongside the likes of the Montauk corpse, Trunko, the Feejee mermaid, and many other monsters of misidentification and fabulous frauds. Moreover, it demonstrates how specimens clearly identified as fakes by their creators (dethcheez, for instance, unequivocally referred to it on his DeviantArt site as a manufactured gaff) are nonetheless all-too-often deemed genuine and erroneously circulated as such online by many other, less discerning people. It also emphasizes, however, just how easy it can often be to expose such specimens' true nature simply by taking time to trace their origins via the internet. This is why it is so surprising to me that such searches are not conducted more often and more vigorously, because this approach would soon eliminate many hoaxes from the database of genuine cryptozoological mysteries.

Incidentally, if you're wondering where the baby chupacabra gaff is right now, some lucky ebay bidder out there may well have the answer—and the specimen—because dethcheez auctioned it on ebay, and the auction ended on April 29, 2007. So if the winning bidder (assuming that it did sell) is reading this chapter, I'd be delighted to hear from you, as it would be good to know something about this fascinating exhibit's current location.

AN ALIEN MOTHMAN, A BURMESE CENTI-SPIDER, AND OTHER MANUFACTURED MONSTERS

Removed entirely from a cryptozoological perspective, there is no question whatsoever that as artistic creations, the gaffs of dethcheez are exceptionally well-produced, compelling, and diverse. Other expertly manufactured examples from this artist's bedazzling menagerie of fantasy fauna include a Jersey devil corpse, a newly-discovered unicorn shark, a vampire mummy's head, a freak multi-legged centi-spider from Burma, a mummified devil turkey head (complete with teeth and a beak horn), a two-tailed twelve-limbed sand scorpion, a snake-headed terrapin, a mummified baby dragon and a baby dragon skeleton, a giant clawed centipede, a mummified baby unicorn, the mounted heads of Peruvian vampire fishes, a mummified winged piglet, a preserved bigfoot finger, plus a vast assortment of shrunken heads, and much much more.

A particular favorite of mine among this eclectic array of the anomalous and bizarre is a preserved "Alien Mothman" (uploaded on September 2, 2006 and viewable in detail at: http://dethcheez. deviantart.com/art/Mummified-Alien-MothMan-2-39107393).   In reality, this eye-catching gaff constitutes a novel variation upon the classic Jenny Haniver/devilfish theme. It consists of a sun-dried ray fish treated with salt and borax, whose large pectoral fins have been converted into wings, with dried, de-furred rabbit feet attached to their tips as clawed hands.

But don't be content with my meager verbal descriptions. Visit dethcheez's online gallery at http://dethcheez.deviantart.com/gallery and experience for yourself his deliciously dark realm of undeniably unnatural history.

COSTA RICAN DODOS ARE A NO-NO

Found only on the Indian Ocean island of Mauritius, the dodo *Raphus cucullatus* is the most iconic extirpated species of modern times. This large flightless pigeon-related bird became extinct in or around 1681, less than a century after its island paradise, previously free of all predators, was invaded and settled by Dutch sailors who killed the dodo for food and disrupted its habitat, aided and abetted by an entourage of hostile introduced species, including rats, domestic cats, dogs, and pigs.

Consequently, a brief video clip purporting to be recent trail-cam footage shot at night by an amateur reptile enthusiast that showed

The author's Portuguese dodo egg-bowl (Karl Shuker)

not only an iguana but also at least two apparent dodos living in the Costa Rican jungle attracted great interest, especially among some cryptozoologists. It had been posted on YouTube (at: https://www.youtube.com/watch?v=dXkikM2M0HE) on March 3, 2015, by someone with the user name Cagey1two3.

Having said that, even to my eyes, untrained as they are in the specifics of cinematographic special effects, it seemed evident that the dodos were merely computer-generated images, and in my view not particularly good ones at that. In addition, I readily noticed that whereas the iguana was always in focus (whether in the foreground or in the background), the supposed dodos were always out of focus (whether in the foreground or in the background). Something was very clearly amiss.

Even so, a few voices seemingly wanted to believe that they were real, despite the many additional, non-cinematographic objections to dodos thriving in Costa Rica that could readily be offered (and indeed, *were* offered by myself and others in various online forums).

How had dodos even come to be in a Costa Rican jungle, many thousands of miles from their Mauritius homeland? There is no documented evidence that dodos were ever taken there. And, of even more pertinence, how had a flightless species from an island originally lacking in predators managed to survive in Costa Rica's predator-replete jungle, which is amply supplied with wild cats, wild dogs, birds of prey, venomous snakes, etc.? And not for just a short time either, but for over 300 years, because if they were real, the videoed dodos must be present-day descendants of specimens transported to Costa Rica and released there prior to their species' extinction on Mauritius during the late 1600s.

True, Costa Rica's jungle is home to a long-established group of large ground-dwelling birds known as tinamous. However, the key difference between dodos and tinamous surviving here is that the latter birds originated in Latin America's tropical forests and grasslands, and they have thus evolved a number of crucial behavioral and morphological adaptations in order to persist in such predator-populated habitats. These include active predator-recognition/avoidance, freeze-motion behavior, cryptic plumage, and although preferring a terrestrial existence nonetheless retaining the ability to take flight if need be (some species even roost in trees).

The dodo, conversely, having evolved in a predator-free homeland, displayed none of these critical survival strategies (which is why they soon became extinct once predators finally arrived on Mauritius). Hence they could not have stayed alive for more than a few days if released into Costa's Rica's jungle, let alone establishing a thriving population here and surviving for more than three centuries, right into the present day. Nor would such distinctive, flightless birds have eluded the many scientific expeditions visiting this locality in modern times. In short, the whole concept of dodos living in Costa Rica was nonsense. And so it proved.

19th-century engraving of a tinamou (public domain)

Ten days later, on March 13, as kindly brought to my notice by Facebook colleague Brian D. Parsons, a second Costa Rican dodo video appeared on YouTube, posted by Leticia Blause (at: https://www.youtube.com/watch?v=fPkB5Hs_f4E), in which a dodo tapped the camera's lens with its beak before displaying a series of captions, written in Portuguese, which stated that although it was too late to save its own species, there were many endangered species out there for which there was still time, if action was taken. The two videos were in fact an advertising campaign for a Brazilian conservation-awareness website, *Quase um Dodo* (*Almost a Dodo*) (at: http://quaseumdodo.com.br/), and one that, thanks to its memorable Costa Rican dodo controversy, had proved to be a brilliantly-executed one, attracting considerable, much-deserved attention to its site and aims.

SPIDERS WITH WINGS WOULD BE TERRIBLE THINGS!

Never in the long and extremely diverse history of spiders—a very significant arachnid order (Araneae) whose lineage dates back more than 300 million years according to the known fossil record—has there ever been a spider with wings. And why should there be? Virtually all spiders display a lifestyle that has no place or need in it for wings, relying upon stealth and ambush to survive and capture their prey, not flamboyant aerial activity like some bizarre eight-legged dragonfly.

Nevertheless, this has not prevented flying spiders from winging their way every so often through both hard-copy and online media reports—to the delight of connoisseurs of the strange and uncanny, and to the despair of hardcore arachnophobes! So here are three of the most entertaining accounts that I have seen, showcasing these faux yet fabulous fliers of the spider kind.

Scientist discovers winged spider

The fake report of a winged spider featuring a photo-manipulated image of an ordinary wolf spider (creator/s unknown)

Case #1 debuted during 2012, when several users of the website Tumblr posted online what initially looked like a bona fide but unidentified newspaper clipping of a supposedly newly-discovered species of winged spider. The clipping consisted of a b/w photograph of the spider in question, entitled "Scientist discovers winged spider," but with no accompanying details concerning it or its discovery. A close look at the photo, however, soon revealed that it was a not-especially efficient exercise in image photo-manipulation. The spider depicted was in fact a common (and wingless!) species of fishing (aka raft) spider belonging to the genus *Dolomedes*.

In addition, as later revealed on the famous hoax-busting *Snopes* website as well as on several others too, the original photograph of it that had subsequently been manipulated by person(s) unknown to yield the winged spider is one that had been snapped on September 23, 2007, in Durham in North Carolina by Will Cook from Duke University in Durham, and had appeared (it still does in fact) on the website *North Carolina Spider Photos* (http://www.carolinanature.com/spiders/bigspider00982.jpg).

On March 10, 2014, the fake clipping and photo were revisited by the website of a UK computer services company, Digital Plumbing, which provided an extensive report about them, including details of how the winged spider, which in this report was unscientifically named Volat-Araneus (it should have been the other way around and italicized, of course, i.e. *Araneus volat*, if the aim was for it to resemble a genuine taxonomic binomial), preyed upon the venomous (and real) false widow spider *Steatoda nobilis*.

However, the report was peppered with clues that it was a hoax, and indeed, halfway through it its (unnamed) writer confessed this openly, explaining that the report's sole purpose had been to attract the attention of readers, who would now, the writer hoped, take note that this website was that of a company offering technology repairs and other services, as detailed in the remainder of the report. In short, Digital Plumbing's report was a very novel marketing ploy, quite possibly the first one ever to utilize a non-existent winged spider to attract potential customers.

Flying spider #2 has only appeared once (to my knowledge)—as an even less convincing photo-manipulated image presented in an extremely brief YouTube video uploaded on October 15, 2013 by Brian Griffin under the title "Have Scientists Discovered a Winged Spider?" (at: https://www.youtube.com/watch?v=pXHnrXbf0zc).

In it, mention is made of the fact that a species called the long-winged kite spider is already known to science. This is true, the species being a forest-dweller known formally as *Gasteracantha versicolor*, which is native to the subtropics and tropics of eastern, central, and southern Africa, as well as Madagascar. However, "long-winged" is something of a misnomer, because its "wings" are not of the membranous, flight-producing variety. Instead, they are a pair of immobile sclerotized spines, borne laterally upon the opisthosoma (abdominal section) of this spider's body in the adult female.

Far older and also far more intriguing than the previous two examples is the third member of this trio of winged wonders—albeit this time a truly grotesque Lovecraftian horror, a veritable cryptic cryptid from the crypts in fact, known as the Italian tomb spider.

I first learnt of this macabre entity courtesy of British cryptozoological archive peruser Richard Muirhead, who sent me an unlabelled review report of an article that had originally appeared in the *Pall Mall Gazette*. Happily, I was soon able to trace the source of this review report—

namely, the *San Francisco Call*, which had published it on November 29, 1896. The report makes such compelling if unnerving reading that I am reproducing it in its entirety below—the first time, as far as I am aware, that it has ever appeared in a cryptozoology book:

San Francisco Call, Volume 80, Number 182, November 29, 1896

ITALY'S TOMB SPIDER

A Thing So Odd That It is Believed to Exist Only in Imagination.

The people of Italy believe in the existence of a wonderful creature which, for the want of a better name, is called the tomb spider. The entomologists know nothing of this queer beast, and declare that it only exists in the fancy of the superstitious persons and those whose curiosity or business makes it necessary for them to explore old ruins, tombs, catacombs, etc. According to the popular account the tomb spider is of a pure white color, has wings like those of a bat, a dozen horrid crooked legs and a body three or four times the size of the largest tropical American tarantula.

The accounts of this queer insect and his out-of-the-way places of abode are by no means common, and on that account the information concerning him which we will be able to give the "curious" is very meager. Any Italian will tell you that such a creature exists, however, and that he is occasionally met with in old mines and caverns, as well as in tombs and subterranean ruins. The London Saturday Review has an article from a correspondent who was present when some Roman workmen unearthed a church of the fifth century. He says: "We were standing by one of the heavy pillars that had originally supported the roof, when something flashed down from the pitchy darkness overhead and paused full in the candle-light beside us, at about a level with our eyes. It was distinctly as visible as a thing could be at a distance of three feet, and appeared to be an insect about half the size of a man's fist, white as wax and with its many long legs gathered in a bunch as it crouched on the stone.

"Our guide had seen, or at least heard of this uncanny insect of ill omen before, but was by no means reconciled to its presence, as his notions proved. He glanced around uncomfortably for a moment and then moved away, we following. It seems really a bit queer, but it is said that the strongest nerves give way in the presence of this insect of such ghostly mien. Even today this uncanny apparition is said to be an unclassified monster—an eternal mystery. When the grave spider is encountered by those opening tombs and vaults it is thought to be a 'sign' of death to one of the workmen or some member of his family." - Pall Mall Gazette

An almost identical account also appeared in another American newspaper, the *Sausalito News*, on January 23, 1897.

What can we say about such a bizarre report? The spider, if indeed we can apply such a name to a creature sporting wings and a dozen legs, is unlike any life form known either upon or beneath the surface of Planet Earth, even if we generously assume that it may be a grossly exaggerated or embroidered description of a pallid form of bat or an exceptionally large moth.

Interestingly, as I documented in my book *The Encyclopaedia of New and Rediscovered Animals* (2012), a dramatically new species of large cavernicolous spider with a pure white abdomen (opithosoma) *was* discovered by science in quite recent times, amid the deeper regions of Koloa Cave on the Hawaiian island of Kauai and in a few others on this same island's southeastern coast, yielding six populations in total. Formally dubbed *Adelocosa anops* in 1973, this spelaean arachnid is an eyeless species of wolf spider, eyes being superfluous as it exhibits an exclusively subterranean lifestyle in which it rarely if ever encounters light.

Although made known to science only fairly recently, this distinctive spider has long been familiar to Kauai's indigenous people, who call it the pe'e pe'e maka'ole. It is easily identified not only by its lack of eyes but also by its long and semi-transparent, orange-colored legs (the normal complement of eight in number), its orange-brown cephalothorax (combined head-and-thorax section), and its ghostly white opisthosoma. Needless to say, however, it does not possess wings!

As for the Italian tomb spider that allegedly does, conversely, during the 19th century, gruesome, highly fanciful yarns of this nature were a popular genre of journalistic reportage, invented purely for entertainment purposes and never meant to be taken seriously, although they sometimes were—especially by the more credulous and less perspicacious of readers. In my opinion, this *San Francisco Call* report from 1896 is clearly a prime example from such a genre.

Nevertheless, I'd still be interested to read the article from the London *Saturday Review* referred to in the latter report (always assuming that such an article does exist), just in case its telling of the tale of Italy's dreaded tomb or grave spider is any less lurid and rather more believable. After all, even an account of a wingless spider sporting only the standard octet of legs typical for its kind but which is unusually large in size, is ghostly-white in color, and exclusively inhabits crypts, catacombs, and other subterranean residences of the deceased would be sufficiently distinct from all recognized spider species to warrant more than passing interest from arachnologists and cryptozoologists alike. So if anyone reading this chapter can trace and send to me a copy of the relevant *Saturday Review* article, I'd very much like to see it—thanks very much!

Of course it is well known that although spiders, being wingless, cannot actively fly, some species can and do practice a type of passive gliding termed ballooning, which is often linked directly to a semi-mysterious phenomenon known as angel hair.

Angel hair is the name given to long, white, gossamer-like filaments that descend earthward sometimes in vast quantities, cloaking and festooning meadows, streets, houses, or anything else that they land upon with shroud-like sheets of their ethereal, silken strands. But what *is* angel hair—and where does it come from? Many eyewitnesses describe angel hair as resembling spider webs, and in most cases this is indeed what it is (but see my book *Dr Shuker's Casebook*, 2008, for some angel hair reports that evidently do *not* involve spider gossamer). Specifically, it is a dense, gauzy aggregation of threads produced by congregations of tiny money spiders (belonging to the taxonomic family Linyphiidae) in order to become airborne via ballooning, as confirmed on several occasions by analysis of samples collected.

Silken threads drawn out of their spinnerets when these spiders face a strong wind are lifted, together with the attached spiders, into the air by the wind and carried aloft, the spiders sometimes travelling great distances before finally gliding back to earth. Once there, they simply abandon their threads, often yielding gossamer-like sheets—angel hair—that are entirely spider-less. This absence can puzzle observers if they have only found angel hair *after* the spiders have already abandoned it. In the case of a very dramatic fall of angel hair in Australia's Southern Tablelands during May 2015, however, the tiny black spiders responsible were still present when it was discovered, having coated an entire town in a thick shroud-like covering of gossamer following a mass ballooning session.

In summary, apart from ballooning spiders, these eight-legged arachnids are reassuringly earthbound, and all are indefatigably wingless—unless you live in Italy and are well-versed in folklore appertaining to grim subterranean realms, and featuring encounters with monstrous creatures that never penetrate up into the light of day, something for which we can all be very thankful, especially if the tomb spider is a typical representative of this shadowy fauna of the catacombs and crypts.

THE MONSTER OF THETIS LAKE—THE TRUTH AT LAST

The term "reptoid" is most closely associated with the reptilian category of extraterrestrial aliens, but it also has a second, less familiar usage, having been applied to a number of extraordinary reptilian humanoid entities reported from modern-day North America and elsewhere. These are also dubbed "Creatures from the Black Lagoon" on account of their outward similarity to the classic horror-movie monster of the same name.

The most famous case featuring a Creature from the Black Lagoon lookalike, however, has recently become the most infamous, due to a shocking yet surprisingly little-publicized revelation. It all began on August 19, 1972, at Thetis Lake, near Victoria, the capital of British Columbia, Canada. This is where two teenage youths, Gordon Pile and Robin Flewellyn, claimed to have seen emerging from the lake a bipedal reptoid covered in silver scales and bearing six razor-sharp spines constituting a central longitudinal ridge running along the top of its head. Moreover, upon seeing the youths, the reptoid lost no time in chasing after them, approaching so closely that it supposedly cut the hand of one of the youths with the spines on its head.

Four days later, during the afternoon of August 23, a similar scenario was reported from the opposite shore of the same lake by two more teenage youths, Russell Van Nice and Michael Gold, who were able to watch the reptoid emerge but without being chased by it this time, because it simply re-entered the water a short time later and vanished. Afterwards, they provided a detailed description of it, which corroborated and added to the description provided by the previous youths. Using this description, a sketch published the following day by a local newspaper depicted the scaly entity with a powerful muscular chest, two three-pronged flippers for feet, clawed humanoid hands, six spines on top of its head, a huge pair of pointed ears, a very large pair of flat fish-like eyes, and an equally piscean mouth.

A bizarre attempt was made by the police to "identify" this entity as nothing more startling than an escaped three-foot-long South American

The Monster
of Thetis
Lake (Richard
Svensson)

tegu lizard—a species notable for *not* walking bipedally, for *not* possessing ears or a spiny crest on its head or flippers for feet, but for possessing a striped body and a very long tail (features conspicuous only by their absence from the eyewitness accounts of the Thetis Lake reptoid). Otherwise, however, nothing more was seen or heard of this lake-dwelling nightmare...until 2009, that is.

This was when Canadian writer-illustrator Daniel Loxton, who edits the *Junior Skeptic* insert section of the highly acclaimed quarterly science-education magazine *Skeptic*, decided to reopen this mystifying case. What spurred him on was his discovery that the very weekend before the first alleged reptoid sighting supposedly occurred at Thetis Lake back in August 1972, a low-budget science-fiction film entitled "Monster From the Surf," featuring a scaly "Creature From the Black Lagoon" type of monster, had been screened not once but twice on local television in this same area of British Columbia. Furthermore, the monster in it perfectly matched the descriptions of the Thetis Lake reptoid that had been given by the teenagers who claimed to have encountered it.

Determined by now to solve this case once and for all, Loxton succeeded in contacting one of the original eyewitnesses, Russell Van Nice (the first time that any investigator had ever done this), who swiftly confessed that their story was a hoax, that they had indeed watched the film on television and had then simply pretended to have seen its monster in real life. True, the testimony of the earlier pair of teenagers has not been exposed as a hoax, but as their description of the Thetis Lake reptoid also corresponds perfectly with that of the monster in the film, it is evident that this reptoid case should no longer be taken seriously.

THE MICHIGAN DOGMAN AND THE GABLE FILMS

For well over a century, there have been reports and claimed sightings in several widely separate localities across North America of bizarre entities said to resemble bipedal wolves or other canids. One such creature is the Beast of Bray Road, named after the quiet country road just outside Elkhorn in Wisconsin, USA, where it was first reported in 1936, and which has been investigated for many years by Linda Godfrey, who has also authored a number of books and articles concerning it.

Another example is the Michigan dogman, originally reported back in 1887 from Michigan's Wexford County. A third such entity has been reported from Vancouver Island in British Columbia, Canada.

Yet although these beings have attracted a great deal of media interest and attention, their stories have singularly failed to receive substantiation from any kind of evidence other than anecdotal—until, that is, the surfacing online in 2007 of the so-called Gable Film.

Earning its title from a name present on a small tag of paper attached to the box containing its reel, this intriguing item consisted of a segment of silent 8-mm film with a total running time of only 3:31 minutes (it can be viewed in full on YouTube at: https://www.youtube.com/watch?v=_jQYbgEZrTA). Seemingly shot in the 1970s judging from the clothes worn by the persons appearing in it and

Do bipedal dogmen genuinely exist? (Richard Svensson)

their hairstyles, most of it is taken up by footage of children riding a snowmobile, which then switches to someone chopping wood, and then to close-up footage of a friendly Alsatian dog in the snow.

In the final segment, however, lasting less than a minute, whoever is shooting the film is now doing so from inside a moving vehicle driving through some snowy terrain, and in the process abruptly records some very brief, shaky footage of an unidentified burly creature moving bipedally nearby in a snow-covered tree-dotted expanse as the vehicle drives by. The film-maker then gets out of the vehicle and pursues the creature for several seconds, but after it turns and faces the film-maker from a distance of around 150 feet away, the creature charges on all fours, and the film-maker flees. Almost at the end of the film (at 3:24), a flash of teeth and a muzzle are recorded by the camera, which then falls to the ground.

A second film surfaced later, dubbed The Gable Film Part 2. Running for just 1:31 minutes and featuring what looks like the same terrain as in the first film, it shows what appears to be the gruesome on-site discovery and investigation by police officers of a ripped-apart corpse that is presumably meant to be the unfortunate maker of the original film (this second film can be viewed online at: https://www.youtube.com/watch?v=CShNKGRY5tw—but beware the grisly footage of the supposed corpse).

Inevitably, the original Gable Film stimulated very appreciable online discussion and dissension as to what it showed and whether it was genuine. Three years later, however, this ongoing mystery experienced the same denouement as that of the Thetis Lake monster—exposed as a hoax. This time, however, the revelation was courtesy of the popular cryptozoological television show *MonsterQuest*. In the final episode of that particular series, screened on March 24, 2010, it unveiled the Gable Film's creator, who was certainly not a ripped-apart corpse but was very much alive and well. His name is Mike Afrusa, and he produced the Gable Film using antique film equipment, film, and props to achieve the period appearance. He was also the dogman itself in the film (wearing a special costume), and the children, dog, and vehicles in the film were his own. The second Gable film was a hoax too.

Afrusa was a long-time fan of a certain song that had been recorded in 1987 by disc jockey Steve Cook at Radio WTCM-FM in Traverse City, Michigan. Entitled "The Legend," Cook's song was inspired by the dogman reports and had played a major part during subsequent years in bringing this entity to widespread attention. Moreover, according to two lengthy posts uploaded by Cook on March 24 and 26, 2010, to his blog *The Dogman Blog* (which can be read in full at: http://michigan-dogman.com/wordpress/?p=176 and http://michigan-dogman.com/wordpress/?p=207), it was none other than Cook himself who had first alerted the *MonsterQuest* team to the fact that the Gable Films were hoaxes—because he had known all along that Afrusa had created them as a tribute to his song. Indeed, even the films' "Gable" tag had originated from the name of Cook's wife's favorite Hollywood actor—Clark Gable.

Perhaps the most ironic aspect of all, however, is that back on September 30, 2007, not long after the original Gable Film had first attracted online interest, Cook and Abrusa had uploaded a webpage uncompromisingly entitled "The Gable Film—Anatomy of an Unintentional Hoax," which they assumed would demonstrate beyond any doubt that the film was a fake—but it didn't! On the contrary, Cook was deluged with emails and telephone calls challenging the webpage's content!

As for the Michigan dogman itself: whether or not such an ostensibly implausible entity truly exists remains an enduring, highly contentious enigma.

EXPOSING THE "DEAD BIGFOOT PHOTO"—THE BEAR FACTS AT LAST!

On November 21, 2006, after having received it from a reader with the user name captiannemo [sic], who claimed to have found it online, veteran American cryptozoologist Craig Woolheater posted on the cryptozoological website *Cryptomundo* the following very intriguing photograph, which has since become popularly known as the "dead bigfoot photo," together with a request for any information available concerning it.

In view of its very striking, tantalizing image, the photo attracted much interest, and was subsequently reposted twice by Loren Coleman on *Cryptomundo* (April 16 and 22, 2009) with further requests for information. It has also been featured on many other websites. Yet although numerous opinions have been aired as to what it depicts (a shot bigfoot, bear, gorilla?) and whether or not it is authentic or photo-manipulated, no conclusive evidence as to its true nature had ever been obtained and presented—until my involvement in the case, that is.

On April 11, 2015, Facebook cryptozoological colleague Tony Nichol brought to my attention the photo reproduced on the next page.

The infamous "dead bigfoot photo" (public domain/photo-manipulated by captiannemo)

With an example of it available for purchase on ebay's USA site, it depicted a shot grizzly bear, photographed in Seward, Alaska, alongside the hunting guide who shot it. The guide's name, as given in white writing running diagonally across the photograph's top-left section, was "C Emswiler," a famous licensed Alaskan hunting guide whose full name was Charles Emswiler (thanks to Facebook friend Bob Deis for informing me of this). The AZO date stamp symbol code on the reverse of the postcard confirmed that the card dates from the time period 1904-1918 (noted by its ebay seller in their auction listing's description of it), and as I could instantly see, its bear photograph was unquestionably the original image from which the bigfoot version had been created by photo-manipulation. After almost a decade, the mystery of the "dead bigfoot photo" was finally solved—except of course for discovering the identity of whoever created it from the vintage bear image.

Vintage picture postcard depicting a hunter and shot Alaskan grizzly bear (purchased on ebay and now owned by Karl Shuker—all rights reserved)

To ensure that it did not become another "missing thunderbird photo," however, I swiftly purchased the example of this picture postcard on ebay, and received it safe and well from its seller on April 24. Perhaps I should also begin scanning ebay for the missing thunderbird photo?!

However, the saga of the "dead bigfoot photo" was not quite over. On April 14, Craig Woolheater announced on *Cryptomundo* that following my revelation of the bear photograph that had served as the source from which the "dead bigfoot photo" had been created (a revelation that I had simultaneously posted on my *ShukerNature* blog and on *Cryptomundo*), he had contacted captiannemo, the *Cryptomundo* user who had sent him the "dead bigfoot photo" back in 2006, and had asked him whether he had created the latter image. In reply, captiannemo confessed that he had indeed created it, and that the copy of the vintage photograph of the bear that he had used for this purpose had appeared in an article on grizzly bear hunting published in an issue of *Field and Stream* from the early 20th century.

So now, not only the source photograph from which the "dead

bigfoot photo" had been created but also a confession regarding its preparation by its creator have been obtained and made public at last. Congratulations to Craig for extracting the confession.

Meanwhile, just when it seemed that the tangled tale of the "dead bigfoot photo" was finally disentangled, another knot of controversy duly presented itself.

On April 15, I received an email from Bill Munns, a much-celebrated cinematographic special-effects expert. His notable contribution to bigfoot investigation is his book *When Roger Met Patty* (2014), in which his extensive analysis of the famous Patterson-Gimlin film purporting to show a female bigfoot swiftly striding into and back out of view at Bluff Creek, California, on October 20, 1967 concluded that the alleged bigfoot (popularly nicknamed Patty, after Patterson) was not a man in a fur suit as many critics believe, but was a bona fide creature.

Bill had now analyzed both the "dead bigfoot photo" and its bear precursor, and to my great surprise he announced in his email to me that in his view not only the "dead bigfoot photo" but also the original bear photograph were hoaxes! He alerted my attention to an illustrated report that he had written, documenting his analysis and containing his reasons for believing both images to be hoaxes, which he had uploaded onto *The Bigfoot Forums* discussion website a short time before emailing me.

Bill's report confirmed that the "dead bigfoot photo" had resulted from not particularly good-quality photo-manipulation of the bear photograph. He then pointed out a number of lighting issues present in the bear photograph that made him believe that it was not a natural outdoors photograph, in spite of its apparent outdoors setting. He also brought to attention what he considered to be suggestions of retouching.

As I am certainly no expert in photographic analysis, and, even if I were, I seriously doubt whether I could match Bill's many years of accumulated experience working in his capacity in the Hollywood film industry, I cannot comment upon the lighting issues that he discusses—other than to wonder whether a photograph known (via the existence of my picture postcard depicting it) to date back almost a century could have been modified so expertly back then. Consequently, I expressed my concern about this in my reply to Bill's email, and in a second email to me, dated April 16, he agreed with me, noting that it would indeed have been a challenge to achieve at that time.

Also, I need to emphasize here that even if Bill's assessment of the bear photograph is accurate and that it is itself a hoax, it does not change anything in relation to its status as the original long-existing image from which the "dead bigfoot photo" was created by captiannemo. This is because, as already noted, my recently-purchased picture postcard containing the bear image confirms the image to be of vintage age, as the postcard's own production dates from the period 1904-1918, i.e. almost a century before the "dead bigfoot photo" appeared on the scene.

Don't forget too that captiannemo stated in his confession of fakery regarding the "dead bigfoot photo" that the copy of the bear photograph that he had used to create the bigfoot version from was one that had appeared in an article on grizzly bear hunting published by the periodical *Field and Stream* during the same time period as the postcard's publication. Speaking of which, it would be good if this particular article could be traced, thereby placing its own existence beyond any shadow of doubt and adding to the postcard's existence a second, independent publication source verifying that the bear photograph dates back at least as far as that early period of the 20th century.

So to anyone reading this account who has access to an early run of *Field and Stream*, if you could check through it and locate the grizzly bear hunting article, I'd greatly welcome its precise publication details (and a scan of it too, if possible).

As far as the "dead bigfoot photo" is concerned, however, all considerations regarding the bear photograph's authenticity are in any case wholly irrelevant (including the obvious fact that because the bear is positioned much closer to the camera than the hunter, the bear looks bigger than it actually is—a familiar optical illusion known as forced perspective, as also documented in Chapter 3 regarding the "giga-gecko"). All that matters is that we know definitely that the "dead bigfoot photo" was created from it via photo-manipulation and is therefore a hoax (with the bear photograph known to have been in existence for almost a century at least).

Having said all of that, later on April 16, I received three more emails from Bill, describing a discovery that he had just made online that had taken him very much by surprise, and leading him to draw a very different conclusion from his initial one regarding the lighting anomalies present in the bear photograph. Here is the first of his three emails:

> I was just doing a bit of research on trick photography, and came on this:
> http://en.wikipedia.org/wiki/Combination_printing
> Apparently putting several image elements into one combination photo dates back to the mid 1800's, and frankly, I find the work quite astonishing, given I know darkroom procedures and can appreciate how painstaking the photo examples shown would be to create.

Needless to say, even if the bear photograph had indeed been manipulated, this discovery now provides a completely different motivation (i.e. from one of simply producing a hoax) for carrying out such an action, as Bill duly acknowledged in his second email:

> Now that I've looked into vintage combination photography printing (the other email), I must wonder how widespread this process actually was for creating impressive photo scenes not conveniently photographed in one setting.

And again, in more detail, in his third email:

> The more I reflect on this "combination printing," the more it seems to have been a respectable form of photographic art, with no intent to deceive, no hoaxing, just a way of creating imagery that could not be easily accomplished in one original photograph. If so, the bear could simply be one such example of recreating a real event that wasn't able to be preserved photographically when it actually occurred.

This seems an eminently sensible conclusion, and in my view it is the most plausible explanation for the various anomalies perceived by Bill in the bear photograph.

Finally, just in case anyone was wondering whether the bear photograph had actually been created from the "dead bigfoot photo" (thereby conveniently ignoring the bear photograph's confirmed very early production date) rather than the other way round, this ridiculous notion was swiftly scuppered as follows by Bill in his report:

> First, the "dead bigfoot" photo can be verified as derived from the

Bear photo because two sections of the Bear body were incorporated into the faked bigfoot shape. And the lower resolution of the bigfoot body photo creates a source/derivative connection that goes one way. Images can be made less sharp, but not more sharp, in the manner shown. Detail could not have been added to the bigfoot photo to achieve the bear photo. But the bear photo detail can easily be reduced to the level of the bigfoot photo.

My grateful thanks to Bill Munns for alerting me to his analysis of the two photos and for discussing this fascinating matter with me at length. Equally, my sincere thanks to Tony Nicol for kindly bringing the bear postcard to my attention, and, in so doing, enabling me to bring the lengthy reign of yet another crypto-pretender to its richly-deserved end.

HOW AN AUSTRALIAN VAMPIRE BECAME THE JERSEY DEVIL

It is rare that a blatant zoological impossibility is not only captured alive but is also placed on public display. Yet during 1909 in New Jersey, USA, this is precisely what happened...sort of.

One of the most bizarre and baffling mystery beasts documented in the annals of cryptozoology must surely be the Jersey devil (aka the Leeds devil, as according to one yarn, it was supposedly the monstrous offspring of Mother Leeds, a reputed New Jersey witch, after she had been impregnated by the devil in 1735). Alleged sightings of this creature date back as far as the 18th century and most commonly occur within an extensive, heavily-forested area of coastal plain stretching across seven counties in southern New Jersey, known as the Pine Barrens. However, there have also been reports emanating from areas far beyond this epicenter—especially during a major "flap" or outbreak of Jersey devil activity that took place during January 1909, and which stretched as far as Pennsylvania, Delaware, and Maryland.

The creature itself, widely blamed for livestock killings and other depredations, has been described in many different ways, but was generally accorded a pair of leathery, bat-like wings, as well as four limbs (with the clawed front limbs much shorter than the cloven-hoofed hind ones). It was also said to be bipedal and was often claimed to sport a horse-like head but bearing a pair of ram's horns, a long forked tail, and flashing red eyes. In some encounters, the eyewitnesses alleged that it had emitted a loud blood-curdling scream.

Needless to say, from a zoological standpoint such a creature is anatomical nonsense, as scientists and other wildlife experts have always steadfastly maintained—thus suggesting that if the Jersey devil is indeed real, it is very different (and far more prosaic) than the evidently much-exaggerated, greatly-embroidered versions that have been claimed by some eyewitnesses down through the ages. Nevertheless, this did not prevent a pair of very canny entrepreneurs from cashing in on the Jersey devil frenzy that took hold during 1909 with what proved to be a very lucrative if entirely fraudulent exhibition purporting to display no less extraordinary an entity than a bona fide, recently-captured, and still very much alive specimen of this sensational(ized) mystery creature.

During the early 20th century, Philadelphia in Pennsylvania was home to T.F. Hopkins's Ninth and Arch Street Museum, but by the onset of 1909 it was suffering from falling attendances. Anxious to reverse this worrying trend, Norman Jefferies, the museum's publicity manager, hatched a mutually beneficial scheme with animal trainer Jacob F. Hope, one that would boost the museum's popularity and also be financially remunerative to Hope in exchange for his assistance.

As concocted by Jefferies and Hope, a certain story was duly made public about how an extraordinary—and extremely bloodthirsty—animal known as an Australian vampire, formerly in Hope's possession, had recently escaped, and how it was this creature that had been responsible for the bizarre Jersey devil sightings reported in the eastern Pennsylvania/ southern New Jersey region during that same time period. Happily, however, according once again to their story, this creature had now been recaptured alive and unharmed, and would be put on display by the museum where it could be seen by one and all, for a viewing fee, of course.

To substantiate this claim, Hope had arranged for a team of a dozen or so specially-hired men who were part of the hoax and acted out the role of trained animal handlers, armed with nets and other implements, to venture forth into a local park one night where the creature was supposedly on the loose, and capture it there. This was duly accomplished when one of the men, perched in a tree, dropped a net over the animal when it passed by underneath. Interested outside

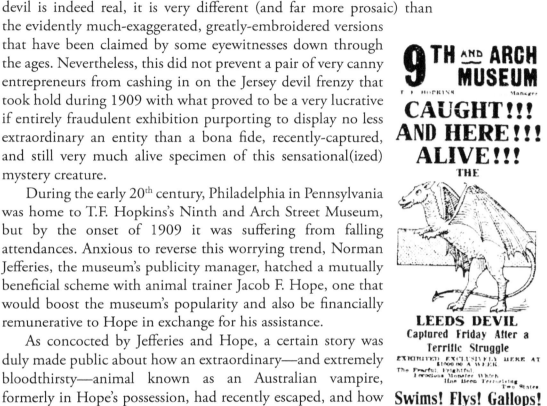

Original advertisement for the Jersey devil exhibition at the Nine and Arch Street Museum in January 1909 (public domain)

observers who may have suspected a fraud, had they directly witnessed what was happening, were kept outside by a fence ringing the entire park, which was also closed to the public at night anyway.

And what was this ferocious "Australian vampire" that they had captured in the park? In reality, it was nothing more alarming than a large male red kangaroo *Macropus rufus* that Jefferies had obtained earlier from a New York animal-dealer colleague and had then "transformed" into the greatly feared Aussie mystery beast. This transformation consisted of Jefferies painting a series of vivid green stripes upon its red fur and attaching to its shoulders a pair of lightweight artificial wings, manufactured from thin bronze and covered with rabbit fur. It had then been secretly taken to the park by its supposed pursuers and quietly released there, after which they had promptly—and publicly—recaptured it.

A massive steel cage placed inside the museum's cellar was then set up by Jefferies as the location for his unique specimen's exhibition. Dimly lit, temporarily hidden from outside view by a dropped curtain, and with a gruesome collection of chewed bones strewn across the floor, this was the melodramatic scene in which the creature would take center stage. Unfortunately, however, the kangaroo did not like its surroundings, and refused to cooperate by putting on any kind of show when tried out by Jefferies in advance of admitting any paying visitors. So he arranged for a boy armed with a long stick bearing a sharp nail at its end to lurk hidden amidst the shadows at the back of the cage.

As soon as the visitors came in and the curtain rose, the boy would poke the kangaroo surreptitiously with the nailed stick, causing it to leap forward shrieking, its false wings flailing, scaring its audience for an instant before the curtain came down. The shocked audience would then leave, the next visitors would enter, the curtain would rise, and the same brief scene would be enacted all over again, and again, and again—because the exhibition proved very popular, generating much-needed increased revenue for the museum, and presumably a handsome payment for Hope too. Only the poor frightened kangaroo, tormented by the nail-embedded stick and confronted by gawping, screaming audiences, gained nothing from the tawdry proceedings.

And how do we know all of this? In 1929, Jefferies publicly confessed to the whole squalid charade.

SETTING THE SEAL ON THE SILVER COELACANTHS? NOT NECESSARILY

In contrast to those preceding it in this chapter, this final case may not be as deceptive as it is commonly deemed to be nowadays. So read its history, then judge for yourself.

Among the greatest ichthyological occurrences of the 20th century was the discovery of not just one but two living, modern-day species of coelacanth—those archaic, armor-plated, lobe-finned fishes hitherto thought to have died out in prehistoric times many millions of years ago. The first species to turn up was *Latimeria chalumnae*, nowadays generally referred to simply as the coelacanth or as the Comoros coelacanth, although some specimens have also been recorded off Madagascar and the eastern and southern coasts of mainland Africa. Indeed, one of these latter vagrants was none other than this species' holotype (type specimen), which was caught alive off South Africa in December 1938. Just under 59 years later, in September 1997, a second species came to scientific attention when, while taking a stroll there on their honeymoon, newly-wed zoologists Mark Erdmann and his wife Arnaz saw a specimen as it was being wheeled on a cart across a fish market in Manado, on the Indonesian island of Sulawesi (Celebes). This new species was subsequently christened *Latimeria menadoensis*, the Indonesian coelacanth, which is brown in color and dappled with shining flecks of gold, whereas the Comoros species is steely-blue with white patches.

The two species' widely separate locations have incited speculation that further living species of coelacanth may perhaps be discovered elsewhere in the world. Moreover, two enigmatic works of art have attracted especial attention as putative support for this prospect.

In 1964, Argentinean chemist Ladislao Reti purchased a small but very beautiful and extremely unusual silver ornament hanging in a village church near Bilbao, on Spain's northern Atlantic coast. Measuring 4 inches long, it was a meticulously sculpted figurine of a fish—but of no ordinary species.

Its scales were thick and armor-like, and its first dorsal fin was borne directly upon its back. But its second dorsal fin, pectoral fins, pelvic fins, and anal fin were all borne upon fleshy lobes, resembling stumpy legs. As for its tail, it bore three fins, the third of which (whose base is just visible in the only known photograph of this ornament) was a tiny lobe sandwiched between the larger upper and lower fins.

Back in 1964 (i.e. 33 years before the discovery of *L. menadoensis*), there was only one known species of living fish that possessed all of these features—in fact, the lobed fins and tripartite tail were unique to it—and that was *L. chalumnae*. Moreover, the silver objet d'art was instantly identifiable as an accurate three-dimensional image of that modern-day coelacanth species. Yet how could its presence in a Spanish church be explained?

Tragically, this remarkable figurine apparently has since been lost (although no details concerning such an event seem to have been made public). However, following examinations of the above-mentioned still-existing photograph of it (reproduced here) by a number of antique silverware experts, the consensus was that it was probably a religious votive that may well have been manufactured during the 17th or 18th century. If correct, this is an astonishing situation—because *L. chalumnae* was not discovered by science until 1938! So how could so accurate a figurine of a coelacanth have been created two centuries or more earlier?

To make matters even more mystifying, in 1965 a second silver ornament in the form of a coelacanth was found, again in Spain. This one was spotted and purchased by Maurice Steinert, who was then a molecular biology student, while browsing through an antiques shop in Toledo, and it is believed to be of similar age to the first.

The surviving photograph of the seemingly now-lost Reti silver coelacanth (International Society of Cryptozoology— c/o J. Richard Greenwell)

Since the discovery of these two silver coelacanths, various solutions have been proposed to explain their anomalous— and ostensibly anachronistic— similarity to *L. chalumnae*. Perhaps they were modeled not upon the living *Latimeria*, but upon comparable fossil coelacanths from the ancient past. Yet if this were true, how can we explain these ornaments' uncannily accurate three-dimensional form? It is hardly likely that their sculptor(s) could have achieved such precise results using only flat fossils as models. Alternatively, their existence may indicate that specimens of *L. chalumnae* had been brought to the Atlantic region prior to the species' scientific discovery in 1938—or even that there is an undiscovered population inhabiting the Atlantic.

However, coelacanth researcher Hans Fricke, from Germany's Max Planck Institute, who was very intrigued by the silver coelacanths, offered a very different answer. While investigating their possible origin, he consulted the world's leading expert on Spanish silver art, José Manuel Cruz-Valdovinos from the Department of Contemporary Art at Madrid University, who opined after examining photos of the two ornaments that they had probably been manufactured in Mexico—while Mexico was still under Spanish colonial rule, thereby explaining their eventual presence in Spain and their Spanish style.

In 2001, Fricke co-authored a scientific paper with Raphael Plante from the Centre of Oceanography in Marseilles, France. In their paper, published by the journal *Environmental Biology of Fishes*, they revealed that they had conducted a close morphological comparison between the existing photograph of the lost Reti silver coelacanth with images of the holotype of *L. chalumnae*. In addition, during May 2000 they had made Steinert's silver coelacanth available for direct examination this time by Cruz-Valdovinos (rather than via mere photos of it as before).

Fricke and Plante concluded that in terms of body shape as well as fin form, position, and orientation, the Reti silver coelacanth appeared so similar to the holotype of *L. chalumnae* that it seemed reasonable to assume that it had been directly modeled upon the latter specimen. As for Steinert's silver coelacanth, Cruz-Valdovinos noted that because, to quote the Fricke-Plante paper, "this artefact lacks style marks, such as peculiar fantasy engravings, typical for baroque silver art pieces of the 16th or 17th century"—and also because of the sharp edges of its body scaling and fins (these edges being of a modern style and contrasting with the smoother edges present on older pieces that have resulted from frequent handling wearing down the edges)—this ornament was probably of recent manufacture. Consequently, he was now discounting his earlier opinion, based only upon photographic evidence, that it was centuries old. And when during May 2000 this specimen was examined independently by A. Jiminez, an expert on silver artifacts and the creation of silver fishes, he opined that it was a typical Spanish folklore art piece and that it was indeed of recent origin, dating from only the past 30-50 years, i.e. c.1950-1970.

As a result of the Fricke-Plante paper's conclusions, it has generally been assumed ever since that the mystery of the silver coelacanths has been solved. However, a careful reading of their paper reveals a couple of glaring contradictions, which do not appear to have been brought to public attention anywhere—until now.

Although most of the Reti silver coelacanth's fins are indeed comparable in shape and orientation to those of the *L. chalumnae* holotype, no mention is made by the authors in their paper that the first dorsal fin of this silver coelacanth looks nothing like that of the *L. chalumnae* holotype. For whereas the latter's first dorsal fin radiates outward from a narrow base and sports a convex upper edge (as is clearly visible in all of the most famous photos of this historic specimen, including the one reproduced below here), the equivalent fin on the Reti silver coelacanth is largely trapezoid in shape, having a very wide base that tapers markedly upwards, with its uppermost edge much less wide and almost straight, bearing a horizontal line of extremely pointed rays running along it. Why such a noticeable discrepancy if this ornament were indeed based wholly upon the *L. chalumnae* holotype, and why no mention of this discrepancy by the authors in their paper?

Equally bemusing is the following outright contradiction. According to the paper's main text, Cruz-Valdovinos referred to a lack of fantasy engravings upon the Steinert silver coelacanth, which is a feature that in turn provided evidence that this ornament was of recent manufacture. Conversely, the caption to Fig. 2 in this same paper (Fig. 2 consisting of two photographs comparing the throat of a *Latimeria* specimen with that of this ornament) stated that the ornament's throat did have fantasy engravings. Moreover, this statement was confirmed by these engravings' readily visible presence in the photograph of the ornament's throat. So how can this contradiction be reconciled?

Photograph of the *L. chalumnae* holotype (public domain)

Following on from that, because the Steinert silver coelacanth does indeed bear fantasy engravings (Cruz-Valdovinos's contrary statement notwithstanding), does this therefore mean that it is not of recent manufacture after all, but actually dates from a much earlier period? If so, this would dramatically pre-date the discovery of *L. chalumnae*, thus re-inciting all of the fevered speculation concerning whether the ornament is evidence for the existence of an undiscovered coelacanth population (or even species) that had been quelled since the publication in 2001 of the Fricke-Plante paper.

Somehow, I think that the seal of a final, unequivocal solution to the riddle of the silver coelacanths has yet to be set.

Chapter 8:
RAT KINGS—A TANGLED TALE OF TANGLED TAILS

And as curiosities go, the rat King seems even more obscure than most, even in its heyday hardly known outside Central Europe. A rat King consists of a number of rats found inextricably tied together by tails in a central knot...The pioneer of cryptozoology, Conrad Gesner, wrote in 1555 of the belief that there were venerable rats who grew in size as they grew more ancient, and were tended by a pack who would steal food and fine velvet for their lord. No giant rat, imprisoned in splendour in a disused barn or cellar, was ever discovered, but as the original plague of black rats spread through Europe and ratting became a secure career for man and dog, the title of rat King was transferred to the occasional discovery of one of these tangled terrors of verminous vermicelli.

— John Michell and Robert J.M. Rickard,
Living Wonders

Among the most bizarre and anomalous of animal phenomena are aggregations of black rats *Rattus rattus* inextricably linked to one another by their tails—which are so thoroughly entangled that the rats have been unable to disentangle themselves and escape. A grotesque, tail-entwined monstrosity of this type is termed a rat king or "roi de rats." This is possibly a corruption of the French "rouet de rats"—"rat wheel"—as the tails when straightened out radiate outwards from the central uniting knot like the spokes of a wheel radiating out from the wheel's central hub. Yet despite centuries of reports and occasional captures, the mystery of how their tails become so intertwined remains unsolved.

MORE THAN FOUR CENTURIES OF RAT KINGS

The earliest currently documented record of a rat king dates from the 1500s, over 450 years ago. It takes the form of a woodcut illustrating a poem in the 4th edition (from 1576) of a monumental emblem book authored by renowned Hungarian scholar-historian Johannes Sambucus, entitled *Emblemata, Cum Aliquot Nummis Antiqui Operis,* and first published in 1564. The poem tells of a rodent-plagued man whose servant discovers seven rats with their tails inextricably tangled together, and this rat king (though not referred to by that term in the poem) is depicted as alive and in some detail within the woodcut. The accuracy of this depiction shows that rat kings were known as far back as the 1500s and suggests that it was based upon a specific example, though no written documentation of this example apparently exists.

Since then, more than 60 specimens have been recorded, spanning the time period 1612-2005, although at least 18 of them are of dubious authenticity (some rat kings have been fraudulently created using dead rats as unusual—and expensive!—souvenirs for the unwary traveler or curio collector). Furthermore, despite being associated with superstitions that their discovery is a portent of the plague and other evils, several rat kings are greatly prized exhibits in various museums.

Intriguingly, most reported rat kings are of German origin, though why this should be is unclear (unless German writers took greater pains to chronicle any such anomalous finds in their own country than writers of other nationalities have done regarding rat kings found in theirs). The single most comprehensive source of rat king information is Martin Hart's book *Rats*, which devotes an entire and very extensive chapter to the subject (once again, moreover, Hart is German, and his book was originally published in German, with an English translation appearing in 1982).

Strasbourg rat king (public domain)

Space considerations obviously prevent me from documenting here every single rat king on record, but a representative selection, including the most dramatic and unusual cases, appear in the following review.

A RAT KING REVIEW

The first on the complete list of specimens currently on record is a nine-rat example that was discovered on March 20, 1612, behind a partition within a loft in Danzig (now Gdansk), Poland, by a local professor. All of the rats were adult, appeared well-fed, and were alive when found. Its details were included in a letter from the professor to a colleague in Basle, Switzerland. This was followed 71 years later by the finding of a six-rat king in Strasbourg on July 4, 1683; these rats were all juveniles.

A very remarkable example was the 18-rat king discovered on July 12 or 13, 1748, by miller Johann Heinrich Jäger at Grossballhausen (also spelt "Gross Ballheiser") in Germany, when it fell from between two stones underneath his mill's cogwheels. Strangely, a famous, beautifully executed copper engraving of this very noteworthy rat king (featured here) depicts it with only nine

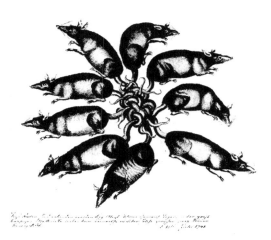

rats present (unless, perhaps, the other nine are hidden beneath the nine portrayed or had been separated from them before the engraving had been executed).

Copper engraving depicting the rat king of Gross Ballheiser (public domain)

Also controversial in terms of its visual portrayal is the rat king discovered in Erfurt, Germany, in 1772. For whereas in his book Hart lists it as a 12-rat king, a detailed engraving of this example dating back to the early 1800s portrays it as containing only 10 rats—but with a very stylized, unnatural-looking knot. Consequently, this picture may have been intended merely as a general rat king representation, rather than as a specific depiction of the Erfurt example.

On January 12, 1774, an amazing 16-rat king was found at a windmill in Lindenau, Germany, with all of its rats still alive. After they had been killed, the king was subsequently displayed in Leipzig, and it proved to be a very popular attraction.

Even more extraordinary was the discovery made during December 1822 at the village of Döllstadt in eastern Germany, when some threshers on a farm found two separate rat kings within a hollow beam in a barn roof

Rat king of Erfurt (public domain)

attic. In both of these kings, the rats were all adult, alive, and apparently healthy. One of the kings consisted of 28 rats, the other consisted of 14 rats. The threshers killed all of them with their threshing flails and then, after great difficulty, the rats in each king were separated. Of particular interest, as originally noted by a forester who witnessed the rats' separation, is that the tail of each freed rat clearly bore the impression of the tails of the other rats in its king, thus demonstrating how tightly their tails had been entwined.

Rat king of Lindenau windmill (public domain)

The most spectacular, monstrous rat king on record, however, was discovered inside the chimney of a miller named Steinbruck in Buchheim, Germany, in May 1828. Incredibly, it contained no less than 32 rats, although they were probably not adults; they were hairless, desiccated, and inescapably bound to one another by their Gordian-knotted tails. This exceptional rat king—indeed, a veritable rat emperor!—is today a much-valued specimen in the Mauritianum, a natural history museum in Altenburg, Germany.

Another preserved rat king can be found in Strasbourg Museum. This is a 10-rat example, whose rats were all juveniles and discovered in a frozen condition under a bale of hay during April 1894 in Dellfeld, Germany. Three of the rats had bite marks, indicating that they may have been bitten by others in the king or had been attacked in their defenseless state by free rats.

Yet another, more recent preserved example is the rat king discovered in 1907 at Ruderhausen, near Germany's Harz Mountains. It resides today in the collections of Göttingen's Zoological Institute. Indeed, this institute may once have possessed a second rat king too, obtained in the very same year, because a number of sources of rat king information list a specimen formerly held at this establishment that was reputedly found in January 1907 in the village of Capelle, near Hamm, in Westphalia, Germany, and brought to scientific attention by the local pastor whose name was Wigger. If these sources are correct, however, it must have since been lost, as no such specimen now exists.

As I learnt in April 2008 from New Zealand correspondent Martin Phillipps, a rat king consisting of eight juveniles has been on public display for several decades in a jar of preserving fluid at New Zealand's Otago Museum. Sometime during the 1930s, it fell down onto the ground, alive, from the rafters of the company shed of Keith Ramsey

Ltd on Birch Street, Otago—followed swiftly by a parent rat that defended them vigorously. The rats' tails were bound together not only by one another but also by horsehair, which is used as nesting material by rats.

In February 1963, a seven-rat king was found by farmer P. van Nijnatten partly concealed under a pile of bean sticks in his barn at Rucphen, in North Brabant, Holland. In the hope of uncovering its secret, after its rats (all adults) had been killed this king was x-rayed, revealing some tail fractures and signs of a callus formation—all indicating that the knot of tails had occurred quite some time ago.

More recent still is the rat king found by some gamebird rearers on April 10, 1986, in the municipality of Mache near Aizenay Vendée, France, and now preserved in alcohol at the natural history museum at Nantes. This king was originally composed of 12 rats, but three subsequently became detached, and it is the resulting nine-rat version that is preserved. The only other preserved French rat king is one that was discovered in November 1889 at Châteaudun and duly presented to its museum where it is still retained today; photographs of it show that it contains six rats, but I have read reports testifying that there were seven originally.

Most recent of all, however, is an Estonian rat king, consisting of 16 rats, nine or so of which were still alive when the king was discovered by farmer Rein Kõiv on January 16, 2005, squeaking loudly on his shed's sandy floor, in the village of Saru. Kõiv killed them all, and on March 10, 2005, the king was taken to the Natural History (Zoological) Museum at the University of Tartu, where it was preserved in alcohol and is now on display. Prior to this, however, a predator had eaten two of the rats and Kõiv had thrown away a third. Two other Estonian rat kings have been reported, both during the 20th century, but neither of them was preserved.

A KNOTTY PROBLEM

A number of explanations for the formation of rat kings have been offered. One of the most popular is that if rats huddle together for warmth in damp or freezing surroundings, with their tails pressed against one another, their tails become sticky and soon adhere to or become frozen against one another, becoming ever more entangled and fixed as the rats thereafter strive to pull free. However, if this were indeed the correct explanation, being such a frequent, commonplace scenario it would surely engender far more examples of rat kings than have been documented so far.

Another notion is that a rat king is actually a single litter whose members' tails became entangled while the rats were still in their mother's uterus. If this were true, however, it seems highly unlikely that they would survive to adulthood, as they would be unable to obtain much food, yoked together in this manner. Yet most rat kings on record feature adult specimens and are often healthy when found. Also, the prospect of the mother being physically capable of giving birth to a single aggregation of united offspring, as opposed to a series of separate offspring, seems very implausible.

Equally intriguing is why all but three of these murine kings feature black rats *Rattus rattus*. There is none involving the much more common brown rat *R. norvegicus*, but this may be due to the fact that the brown rat's tail is shorter, thicker, and less flexible than that of the black rat. Indeed, the only rat exception to the black rat rule is an Indonesian king consisting of ten young Asian field rats *R. brevicaudatus*, discovered on March 23, 1918 in Buitenzorg (aka Bogor), Java. In addition, a single king composed of house mice *Mus musculus* has been recorded (documented in a Russian book dealing with mice and rats). So too has one field mouse king, consisting of several juvenile long-tailed field mice *Apodemus sylvaticus*, found at Holstein, Germany, in April 1929.

SQUIRREL KINGS
However, not all rodent kings involve rats or mice. Several squirrel kings have also been recorded, even though the concept of bushy-tailed squirrels becoming entwined together in this way seems even more incongruous than that of rats. A seven-squirrel king was discovered in a South Carolina zoo on December 31, 1951. Tragically, however, two of its members were dead, a third was dying, and its four other members could only be separated by cutting off their tails above the knot. Curiously, two other squirrel kings have been found here over the years, and all during cold snowy weather, implying that the squirrels had huddled together for warmth.

A king containing six young squirrels was spotted by schoolgirl Crystal Cresseveur in a hedge outside her home in Easton, Pennsylvania, USA, on September 24, 1989. Although they were eventually rescued, their tails could not be disentangled. A five-squirrel king (two members of which were albinos) fell out of a tree by Reisterstown Elementary

School in Baltimore, Maryland, on September 18, 1991, but these were successfully separated as their tails were linked to one another only indirectly, with sticky tree sap, tangled hair, and nest debris. In Europe, at least two kings of red squirrels *Sciurus vulgaris* have been recorded—one in August 1921, the other on October 20, 1951 (which was later preserved).

In July 1997, what looked at first like a huge hairy spider was spotted under a tree at Brantford, in Ontario, Canada. However, a closer look revealed that it was a squirrel king, composed of five young squirrels whose tails were braided together right up to their bases. They were taken to a vet, Cathy Séguin, who freed them, but she feared that the loss of circulation suffered by their tails would result in part of each tail dying.

CAT KINGS

Analogous to rat kings, mouse kings, and squirrel kings is the even more obscure phenomenon of cat kings. Here, however, it is the umbilical cords of kittens in newborn litters that are tangled and intertwined, not their tails. Such curiosities are extremely rare, but at least a dozen have been documented in the scientific literature. Perhaps the best known is the cat king recorded in October 1937, at Rennes, in Brittany, northwestern France. It consisted of a litter of eight small kittens, seven of which (males and females) were held closely together via their entangled umbilical cords. Indeed, the entanglement was so complex that even the left hind legs of two of the kittens had become bound up to one another. The kittens in this cat king were discovered dead, but it is not known whether their cords' complicated intertwining had occurred before the litter's birth or afterwards.

A cat king recorded in 1841 consisted of five unborn kittens whose

Engraving of the 1841 cat king (public domain)

entwined umbilical cords were still fixed to their mother's placenta. This shows that such entanglement can indeed occur while the kittens are still in the womb. The earliest cat king known to me is a five-kitten specimen from 1683, found in Strasbourg, France.

As for rat kings, the mystery of how they are formed remains exactly that—an unexplained anomaly whose reality has spanned centuries but whose secret still awaits disclosure. Some skeptics have cynically suggested that the phenomenon is simply a hoax, that the knotting together of the tails of rats in a king is clearly the result of deliberate human activity. Bearing in mind, however, that most of the rat kings on record were discovered when their rats were very much alive and often still in good health (albeit hungry and frightened), and that the tail knots were exceedingly complex, all that I can say in response to these skeptics is that they have certainly never attempted to tie the tails together of several (not to mention 16 or 28!) live, healthy rats. If they were ever brave (or foolish) enough to try to do so, their cynicism would very swiftly vanish—along, quite possibly, with several of their fingers!

Chapter 9:
THE VEGETABLE LAMB OF TARTARY—
HALF PLANT, HALF ANIMAL, WHOLLY AMAZING!

Feeding on grass, and th'airy moisture licking
Such as those Borometz of Scythia bred
Of slender seeds, and with green fodder fed;
Although their bodies, noses, mouths and eyes,
Of new-yeaned lambs have full the form and guise,
And should be very lambs, save that for foot
Within the ground they fix a living root
Which at their navel grows; and dies that day
That they have browsed the neighbouring grass away.
   — Guillaume de Salustre du Bartas, *La Semaine*

For many centuries, naturalists seriously believed that a small fleecy creature originated from a truly extraordinary plant's fruit and was therefore a unique fusion of zoology and botany. Hence its name—the vegetable lamb.

The most extensive modern-day documentation of the vegetable lamb, also known as the barometz or borometz, can be found in Jan Bondeson's fascinating book *The Feejee Mermaid and Other Essays in Natural and Unnatural History* (1999). In it, he reveals that lore relating to lambs supposedly growing out of the ground dates back many centuries in China. Moreover, the earliest known mention of such a creature anywhere appears in a Jewish text from 436 AD entitled the *Talmud Ierosolimitanum*, or *Jerusalem Talmud*, written by Rabbi Jochanan, which refers to the yeduah, a lamb-like beast that sprouted from the ground attached to a plant stem. However, it was not until the 14th century and the publication of a certain English nobleman's extraordinary travelogue that this bizarre plant-animal first attracted appreciable Western attention, after which it swiftly became a staple inclusion in any self-respecting bestiary.

The travelogue in question chronicled the astounding voyages of Sir John Mandeville around the then-known world, in which he claimed to have personally observed all manner of incredible and highly implausible creatures, including the vegetable lamb. He supposedly encountered this entity during his sojourn in a region of Tartary (a name used at that time for much of northern and central Asia) that nowadays constitutes China. Here is what he wrote about it:

> There grows there a kind of fruit as big as gourds, and when it
> is ripe men open it and find inside an animal of flesh and blood and
> bone, like a little lamb without wool. And the people of that land eat
> the animal, and the fruit too. It is a great marvel.

Vegetable lamb
as portrayed in
Mandeville's
travelogue, 14th
century (public
domain)

In later centuries, it was revealed that Mandeville had never existed and that his travelogue was a clever hoax, quite probably executed by a 14th-century Benedictine monk of Flemish extraction named Jan de Langhe, ingeniously incorporating and interpolating tracts extracted from several earlier works penned by real writers (a medieval Italian Franciscan friar and explorer named Odoric of Pordenone in the case of this travelogue's vegetable lamb information). But by then, the fictional Mandeville's equally fictitious coverage of the vegetable lamb had firmly taken root, in every sense, firing both the imagination of Western naturalists anxious to see for themselves this true wonder of Creation and the inspiration of Western artists including depictions of it in religious illustrations. Perhaps the best example of the latter is the very detailed, ornate frontispiece plate included in English herbalist John Parkinson's monumental treatise *Paradisi in Sole Paradisus Terrestris* (1629), in which a vegetable lamb can be perceived just behind Adam in the Garden of Eden.

During the mid-16th century, an equally influential account of the vegetable lamb appeared, this time penned by the celebrated scholar-diplomat Baron Sigismund von Herberstein (1486-1566), who had twice been the German emperor's ambassador at the Court of Muscovy (a Russian principality centered on Moscow). In his account, published in 1549 within his magnum opus *Notes on Muscovite Affairs*, he added several important details derived from information passed on to him by a number of different Russian sources.

In contrast to the Mandeville description claiming that it lacked wool, the Baron's account stated that the vegetable lamb possessed not only a normal lamb's head with eyes and ears, but also a normal lamb's woolly fleece. Its tiny limbs even sported hooves, though these were exceedingly delicate as they were apparently composed merely of compressed hairs, not the hard horny substance of real lambs' hooves. The lamb was permanently attached to a long stem, comparable to an umbilical cord, which grew vertically to a height of approximately 2.5 feet, thus suspending the lamb high above the ground, but it could apparently use its weight to bend the stem downwards, thereby enabling it to stand and walk upon the ground, and also to graze upon any grass or foliage that was within its reach.

Unfortunately for the vegetable lamb, however, as documented by botanist and fervent barometz believer Claude Duret in his *Histoire Admirable des Plantes et Herbes Esmerueillables et Miraculeuses en Nature* (1605), its flesh was very palatable (said by those who had eaten it to taste like crab meat) and its blood resembled honey. Consequently, it attracted particular gastronomic attention not only from humans but also from marauding packs of wolves, against which the little lamb had no defense. It could not even flee them, as it was irrevocably attached to its stem, and so was invariably torn apart and devoured by its ravaging attackers. Nor was that the only tragic fate that regularly befell this poor creature. Due again to its permanent tethering via its stem, once the lamb had eaten all of the grass and other vegetation within its reach it was doomed to starve to death, after which its plant progenitor died too.

Yet although such tales and accounts made absorbing reading, even in that pre-scientific age scholars still sought physical evidence to corroborate them whenever possible—but what physical evidence existed to confirm the reality of the vegetable lamb? According to the Tartars, they utilized this creature's fine wool as padding for the caps that they wore on their shaven heads at night for warmth, and also—of particular excitement to Western naturalists—some Muscovites claimed that the Tartars would occasionally sell entire vegetable lamb skins, albeit only for inordinately high prices.

As recorded by Jan Bondeson in his own comprehensive barometz writings, one person who was aware of such claims was Sir Richard Lea, who in 1570 had been appointed the ambassador of England's Queen Elizabeth I to the court of the Russian Tsar, Ivan IV ("The Terrible"). Moreover, he actually succeeded in obtaining a coat lined with vegetable lamb skins, after trading for it with the tsar an exquisite grinding mortar hewn from a magnificent piece of agate. Upon his death in 1609, Sir Richard bequeathed this zoo-botanical (or phyto-zoological?) treasure for safekeeping and study to none other than Oxford's now world-famous Bodleian Library, which had been founded during that same period of time by Sir Thomas Bodley (1545-1613). Sadly, however, his expectation was not met, as the coat was simply left to deteriorate in condition inside Sir Thomas's own closet. Despite attempts to repair and renovate it during the 1630s and 1640s, it was probably discarded not long afterwards, because by the end of that century its whereabouts were no longer known and have never been ascertained since.

Notwithstanding this very regrettable loss (although 17th-century German naturalist Engelbert Kaempfer revealed that other such artifacts sold by Tartars were actually derived from the skins of unborn Astrakhan lambs), several entire preserved vegetable lambs have also been formally documented. One, measuring just over one foot long, resembled a four-legged wooden branch covered in a shining dark-yellow fleece.

It had originally been purchased from an Indian merchant by a Mr. Buckley, and in 1698 it was exhibited at the Royal Society of London by the Society's Secretary, Sir Hans Sloane (whose own extremely substantial collection of artifacts became the foundation of the British Museum after he bequeathed them to the nation).

Engraving of Buckley's vegetable lamb, exhibited by Sir Hans Sloane (public domain)

Moreover, Sloane exhibited a second preserved vegetable lamb at the Royal Society in 1725, this specimen originating in Russia and belonging to German physician Johann P. Breyn. As Jan Bondeson has so aptly commented, however, it looked more like a fox terrier than a lamb!

Sadly, both of those specimens are now lost, but at least two others do still exist. One of them is a prize exhibit at the Garden Museum in Lambeth, London, which I specifically visited on February 6, 2015 in order to see it. When I arrived, however, I was sad to discover that it was not presently on display, but after the museum's exhibitions curator, Emily Fuggle, learnt of my interest in mysterious and mythological creatures she very kindly treated me to a private viewing of their celebrated specimen, currently residing in the museum's storage center. Standing in silent dignity, a mute and motionless marvel from a long-bygone age, the vegetable lamb of Lambeth peered ever outwards through the large glass dome inside which it was detained.

This memorable specimen, probably created during the mid-1800s, was an unexpected but very welcome donation to the museum some years ago from a Cambridgeshire doctor whose family had owned it for over 150 years, but Emily informed me that it is now too fragile and vulnerable to the effects of light and photography from which its antiquarian glass cupola can no longer shield it adequately for it to be placed on public display at present. Happily, however, there are plans for this unique wonder to return on show at the museum as a permanent exhibit in 2016, housed inside a special new case affording it full protection, so I am already looking forward to a return visit there to see it again.

Engraving of Johann Breyn's terrier-like vegetable lamb (public domain)

The second preserved vegetable lamb specimen, which has resided inside its very own chest of drawers for over 200 years, is maintained in the storage center of London's Natural History Museum, having only been placed on display once—briefly, in 1934—during modern times. An engraving of it was prepared during the 18th century by John and Andrew Rymsdyk, and appears in their *Museum Britannicum* (1778).

Needless to say, vegetable lambs do not, could not, exist, and never have existed—they are nothing more than an exotic, imaginative fable from the Middle Ages. So how can the preserved specimens be explained—what exactly *are* they?

After examining the two examples that he exhibited at the Royal Society, Sloane had no doubts whatsoever concerning their identity. Both of them were nothing more than the inverted, hairy rhizome or

The Natural History Museum's vegetable lamb, depicted in an engraving from 1778 (public domain)

Vegetable lambs depicted in artificially modified, pseudo-zoological form on left, and in natural, fern-bearing form on right, from *Svenska Familj-Journalen*, vol. 18, 1879 (public domain)

rootstock of some form of large fern, whose roots had been removed, and four of whose frond stems had been retained but carefully shaved and modified to resemble slender legs. And the only reason why this correct identification had not been readily recognized earlier is that the fern species in question did not begin to be widely introduced into Europe from its native habitat in China and the Malayan Peninsula until the 1800s. A very large arborescent tree fern that grows up to three feet in height and whose fronds can reach 10 feet in length when fully mature, it is now commonly known as the woolly fern and has been scientifically dubbed *Cibotium barometz*—both names commemorating its link to the vegetable lamb legend.

All that remains to be answered, therefore, is how the myth of the vegetable lamb arose in the first place. In a concise book devoted to this legendary entity, published in 1887 and entitled *The Vegetable Lamb of Tartary*, Brighton Aquarium naturalist Henry Lee proposed that it was inspired by the cotton plant *Gossypium herbaceum*, whose white clumps of fleecy cotton fibers surrounding the plant's seeds (revealed when its ripe seed pods burst during warm weather) superficially resemble tiny lambs attached to stems. This hypothesis has been supported by a number of subsequent researchers. However, as Jan Bondeson has tellingly pointed out, the cotton plant had been a familiar, widely

utilized species in Europe for
centuries, and for even longer in
China, yet with no suggestion
anywhere on record of any
myths or fables linking it to the
production of lambs. So although
initially appealing, Lee's proposal
is definitely lacking in material
support.

Consequently, although we
know unquestionably that the
vegetable lamb as a biological
reality is an impossible concept,
the riddle of how belief in this
most fantastic of fantasy life-
forms began, becoming an
enduring myth in China, the
Middle East, and thence Europe,
still lacks a convincing answer
even today.

Vegetable lamb as
depicted in Henry
Lee's book, *The
Vegetable Lamb
of Tartary*, 1887
(public domain)

Chapter 10:
OARFISH ORIGINS—AND A VERY (UN?)LIKELY SEA SERPENT

If cryptozoology is ever going to hit pay dirt, the jackpot is most likely to be marine. Even inshore waters are a mystery, and it is the height of hubris to think we've uncovered all the big surprises. It's certainly conceivable—perhaps not likely, but conceivable—that one or more large unknown species that fit the old "sea serpent" mold are hiding out there, too, ready to shock and delight us one of these days.

— Scott Weidensaul,
The Ghost With Trembling Wings

One of the world's most extraordinary, enigmatic, and famously elusive animals, the giant oarfish *Regalecus glesne* is truly spectacular. What other fish can boast a silver-skinned, scaleless, laterally-compressed, ribbon-like body of illusively serpentiform appearance known to measure over 30 feet long (and with plausible if unconfirmed lengths of up to 50 feet also documented—see below); a blood-red erectile crest composed of the first few greatly-elongated rays of the dorsal fin and memorably compared to a Native American's head-dress by one of the world's most famous science-fiction writers; an equally erythristic but shorter-rayed remaining dorsal fin running the entire length of its body; a horse-like head with protrusible toothless jaws; and a pair of very long, oar-shaped pelvic fins that earn this singular species its most frequently-used common name?

Engraving of a giant oarfish underwater, from *The Royal Natural History* (1896), edited by Richard Lydekker (public domain)

The giant oarfish is the world's longest species of bony fish (Osteichthyes), but the question asked more than any other about this species is: Just how long *is* it? The most authoritative answer is

as follows, quoted from Mark Carwardine's standard work on animal
superlatives, published in 2007 by London's Natural History Museum
and duly entitled *Natural History Museum Animal Records*, which is also
the data source cited by *Guinness World Records* (formerly *The Guinness
Book of Records*):

> A specimen seen swimming off Asbury Park, New Jersey, USA,
> by a team of scientists from the Sandy Hook Marine Laboratory
> on 18 July 1963, was estimated to measure 15 m (50 ft) in length.
> Although this is purely an estimate, it is noteworthy because it was
> seen by experienced observers who, at the time, were aboard the 26
> m (85 ft) research vessel *Challenger*, which gave them a yard stick for
> measuring the fish's length. With regard to scientifically measured
> records, there are a number of oarfish exceeding 7 m (23 ft) in length;
> for example, in 1885, a specimen 7.6 m (25 ft) long, weighing 272
> kg (600 lb), was caught by fishermen off Pemaquid Point, Maine,
> USA.

One of the scientists aboard *Challenger* when it had its close
encounter with that mega-large giant oarfish in 1963 was Lionel A.
Walford from the Sandy Hook Marine Laboratory of the Bureau of
Sports Fisheries and Wildlife. In a subsequent interview, Walford
evocatively recalled that it "resembled a transparent sea monster. It
looked like so much jelly. I could see no bones, and no eyes or mouth.
But there it was, undulating along, looking as if it were made of fluid
glass."

A near-legendary yet globally-distributed inhabitant of tropical
and temperate mesopelagic waters from 660 to 3300 feet in depth,
the giant oarfish is a member of the taxonomic order Lampriformes
(aka Lampridiformes), whose other members include the ribbonfishes,
dealfishes, opahs or moonfishes, crestfishes and bandfishes, taper-tails,
thread-tails, and velifers.

Together with the oarfishes, they are collectively known as lamprids
and constitute some 20 species in seven families. Most lamprids possess
long, ribbon-shaped (taeniform) bodies, the remainder (most notably
the opahs) are rounded, deep-bodied (bathysome); all are laterally
flattened, and most have bright red fins, and often a very lengthy dorsal
fin.

The giant oarfish is the only member of its genus, *Regalecus*, and, with
a single exception, is the only member of its entire taxonomic family,

Regalecidae. That lone exception is the streamer fish *Agrostichthys parkeri*—a lesser-known species superficially similar in basic appearance to *Regalecus* but much shorter (no more than 10 feet long), and also possessing far fewer gill-rakers (8-10, as compared with 40-58 by the giant oarfish).

Interestingly, the streamer fish is apparently electrogenic, because people handling specimens of it sometimes claim to have experienced a very mild electric shock. However, no such effect has apparently been reported in relation to the giant oarfish (which in view of its much greater length is probably just as well!).

The streamer fish was formally described and named in 1904, when it was housed with the giant oarfish in the genus *Regalecus* as *R. parkeri*, but in 1924 it was reassigned to a separate, newly created genus, *Agrostichthys*, in which it remains to this day. This mysterious species is currently known only from seven specimens, all collected in southern oceans.

Moreover, due to its deep pelagic existence, the giant oarfish is also notably under-represented by physical specimens (despite its far bigger size), with most of those that *have* been documented consisting of specimens that have been beached after storms or found dying or dead in coastal shallows.

Engraving of a North Pacific crestfish *Lophotus capellei* (also known as the unicorn fish for obvious reasons), from *The Royal Natural History* (1896), edited by Richard Lydekker (public domain)

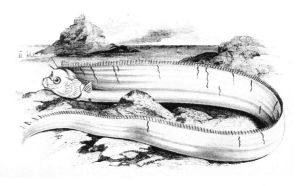

Yet regardless of its evanescence, *Regalecus* has been known to science for a much greater time-span than *Agrostichthys*, having been officially described and named as long ago as the second half of the 18th century by the Norwegian biologist Pedr (aka Peter) Ascanius (1723-1803), who had been a student of the great Linnaeus.

Intrigued to read this historic scientific account, I spent quite some time seeking it online, but finally succeeded in unearthing a copy of

Engraving of a beached giant oarfish, from *A History of the Fishes of the British Islands*, 1862-1866 (public domain)

the description in question. Just a few lines long, it was published on page 5 in Part 2 of Ascanius's great work—*Icones Rerum Naturalium, ou Figures Enluminées d'Histoire Naturelle du Nord*, written primarily in French, but with species descriptions written in Latin. Part 2 was published in Copenhagen in 1772. It also included an illustration of this dramatic species' type specimen.

Tab. XI.

LE ROI DES HARENGS.

La grandeur de ce poisson est de 10 à 12 pieds. Le seul endroit où il a été trouvé est à Glesvær près de Bergen. On l'appelle *Sild-Konge*, *Sildtus*. Ce genre est nouveau & l'espece est unique. On en donnera l'histoire détaillée dans les *Memoires de l'Académie des Sciences de Copenhague* pour l'an. 1770.

Regalecus glesne, cirris pectoralibus longissimis; Pinnæ dorsalis radii priores soluti subspinosi.

B. 4 — 5. D. 120. P. 10. V. 1. A. 0. C. —

Regalecus. *Caput laminis osseis tectum; Maxilla prominentes; Dentes subulati; Membrana branchiostega radiis IV — V.*

Pinnæ ventrales filiformes, analis nulla; corpus ensiforme lineis asperiusculis. Norv. *Sildtus.*

Ascanius's description and illustration of the type specimen of the giant oarfish *Regalecus glesne* Ascanius 1772 (public domain)

As seen in his description, Ascanius formally named the giant oarfish *Regalecus glesne*, which is still accepted as its official binomial name, although during the years that have followed Ascanius's account, many other binomials have been applied to it, all of which are now deemed to be junior synonyms. Having said that, some researchers deem *Regalecus russelii*, named by the eminent French zoologist Baron Georges Cuvier in 1816, to be a valid second species (in particular, Tyson R. Roberts in his comprehensive monograph *Systematics, Biology, and Distribution of the Species of the Oceanic Oarfish Genus Regalecus (Teleostei, Lampridiformes, Regalecidae)*, 2012). Others, however, consider it to be conspecific with *R. glesne*. Ditto for *Regalecus pacificus*, named in 1878; and *Regalecus kinoi*, named in 1991.

Regalecus signifies kinship to a king and is derived from the giant oarfish's popular alternative name, "king-of-the-herrings" (the name utilized as a common name for it by Ascanius in his description). That in turn is derived from a longstanding folk tradition that this gigantic species leads shoals of herrings to their spawning grounds. A comparable folk-belief among the Macah people west of Canada's Strait of Juan de Fuca has earned a related fish, *Trachipterus altivelis*, a species of ribbonfish, the common name "king-of-the-salmon."

The giant oarfish's specific name, *glesne*, derives from the name of a farm at Glesvaer (aka Glesnaes), near to the major Norwegian city of Bergen, where this species' type specimen was found. As for the name "oarfish," this originates from an early false assumption that this species swims by circular, rowing movements of its oar-shaped pelvic fins (scientists nowadays believe that these unusual fins are used for taste detection).

In reality, this elongate species' swimming movements are much more intriguing, and diverse, as it can swim holding its body horizontally and also holding it vertically. In horizontal mode, it moves by undulating its body-length dorsal fin while keeping its body straight (a mode of locomotion known as amiiform swimming—named after a primitive, unrelated North American freshwater fish called the bowfin *Amia calva*, whose own lengthy dorsal fin performs the same undulatory activity for swimming purposes). In July 2008, while kayaking in Baja California, Mexico, on a trip organized by Un-Cruise Adventures, guests filmed two giant oarfishes exhibiting amiiform swimming in shallow water. The oarfishes were each around 15 feet long, and an excellent-quality video filmed of them by one of the guests can be viewed on YouTube (at: https://www.youtube.com/watch?v=IoDqG0syBFE).

As he exclusively documented in the June 1997 of the British magazine *BBC Wildlife*, during a recent dive off Nassau in the Bahamas Brian Skerry was fortunate enough not only to encounter a living giant oarfish at close range but also to photograph it—and he was amazed to observe it holding its long thin body not horizontally but totally upright and perfectly rigid, with its pelvic rays splayed out to its sides to yield a cruciform outline, while seemingly propelling itself entirely via movements of its dorsal fin. Until then, no one had suspected that this serpentine species could orient itself and move through the water in a perpendicular fashion. Ichthyologists now believe that the giant oarfish specifically adopts this vertical or columnar stance when searching for prey. In 2010, Serpent Project scientists videoed via ROV (remote-operated vehicle) a very big specimen of the giant oarfish, measuring between 16 and 32 feet long, swimming underwater both horizontally and vertically in the Gulf of Mexico (their video, as posted by Discovery, can be viewed online at: https://www.youtube.com/watch?v=lvRqqwBoyx8). It is the first film of this species swimming in its natural, mesopelagic zone habitat, rather than in shallow water.

Any self-respecting cryptozoological enthusiast will tell you that the giant oarfish is a popular mainstream explanation for sightings and reports of at least some alleged sea serpents—and after all, with its enormous length and extremely elongate form, this is surely little wonder. Although it is normally a mesopelagic species, existing at depths from 660-3300 feet, occasionally a specimen will enter shallower, coastal waters, and a number of strandings have been reported through the years.

For instance, on January 22, 1860 (not 1880, as some accounts claim), a dying *Regalecus* measuring 16 feet 7 inches long but less than 1 foot wide was discovered washed ashore at Hungary Bay on Bermuda's Hamilton Island by George Trimingham and a relative as they strolled along the beach there, and was labeled as a dead sea serpent by a Captain Hawtaigne from the British Army's 39th Regiment of Foot in a letter published later that same year in *The Zoologist* (even though his description of it left no doubt whatsoever that it was a giant oarfish). Happily, the creature's true identity was swiftly confirmed when its carcass was examined thoroughly soon afterwards by Bermuda-based naturalist J. Matthew Jones.

Moreover, in a letter to *The Times* newspaper of London, published on June 15, 1877, British zoologist Andrew Smith voiced what remains today a popular consensus among the scientific community when he confidently asserted:

> I am, as a zoologist, fully convinced that very many of the reported appearances of sea-serpents are explicable on the supposition that giant tape-fish [i.e. giant oarfishes]—of the existence of which no reasonable doubt can be entertained—have been seen.

Consequently, it may come as something of a surprise to discover that Bernard Heuvelmans, the Father of Cryptozoology himself, no less, was scathing about the idea of giant oarfishes being mistaken for sea serpents in his mighty tome *In the Wake of the Sea-Serpents* (1968). He pointed out that this species' very large, unique, bright-red crest would readily identify it for what it truly was—a giant oarfish, thereby unequivocally differentiating it from any serpentiform marine cryptid.

Heuvelmans also claimed that the biggest giant oarfish specimen ever accurately measured was only just over 21 feet long. He discounted all reports of longer specimens as exaggerations, adding uncompromisingly: "It seems that the only reason why there has been an attempt to stretch the maximum size of the [giant] oarfish, is in order to explain the sea-serpent by an animal known to science."

Engraving of Bermuda's Hungary Bay giant oarfish, sketched by W.D. Munro for the March 3, 1860, issue of *Harper's Weekly* (public domain)

These seem harsh criticisms, but I must point out that they were written before confirmed specimens exceeding 21 feet were discovered (except, that is, for the 25-foot Pemaquid Point individual of 1885, which, oddly, Heuvelmans did not mention in his book). They were also written before films of living oarfishes were obtained—films showing that the vivid red crest is actually nowhere near as conspicuous when the fish is swimming as Heuvelmans had apparently assumed it would be.

Moreover, if observers who are not familiar with this species should see a giant oarfish when it is swimming in horizontal, amiiform mode, or even if found stranded ashore (like the Bermuda specimen), it is easy to understand why they might indeed be wondering if they had encountered a veritable sea serpent from the deep—possibly even a maned one. For the giant oarfish's long, low dorsal fin might well explain sightings of elongate sea serpents sporting manes.

One type of sea monster that Heuvelmans did feel certain was linked directly to the giant oarfish, conversely, was a specific type of marine serpent dragon that featured in a famous story from classical Greek mythology.

During the Trojan War, Laocoön, a priest of Poseidon, voiced his suspicion that the wooden horse of Troy given by the Greeks was some sort of trick, not to be trusted, and begged for it to be destroyed. In response, the Greeks' divine supporter, the goddess Athena, sent two enormous limbless sea dragons with blood-red crests through the waters until they reached Laocoön, whereupon they emerged and killed him, as well as his two sons.

Heuvelmans's linking of these crested sea dragons with the giant oarfish seems reasonable, as the story may well have been inspired at least in part by a Mediterranean stranding of one or more giant oarfishes, whose striking appearance would no doubt have stayed long in the memories of those who witnessed them.

"Laocoön and His
Sons"—marble
statue, c.200 BC
(public domain)

Nor are sea serpents and marine dragons the only legendary beasts that have been associated with the giant oarfish. So too have Asia's ancient snake deities, the nagas, as I noted in my book *Dragons in Zoology, Cryptozoology, and Culture* (2013):

Allegedly seized from the Mekong River by the American Army in Laos on 27 June 1973 during the Vietnam War, a supposed queen naga or nagini is depicted in a famous much-reproduced photograph that is often seen displayed as a curio in tourist bars, restaurants, markets, and guest-houses around Thailand. However, the creature in question is visibly recognisable as a dead [giant] oarfish, held up for display by a number of men.

Moreover, it is now known that this oarfish specimen, measuring 25.5 ft long, was actually found not in Asia at all, but off the coast of Coronado Island, near San Diego, California, by some US Navy SEAL trainees in late 1996, and those are the men who are holding it.

There are also two little-known Icelandic sea monsters that may have been inspired by reports of the giant oarfish, judging from their bright red dorsal crests. For although this species is not generally found in Arctic waters, it is known from Scandinavian coasts further south (its holotype being one notable example). These monsters are the red-maned hrosshvalur or horse-whale and the aptly named raudkembingur or red-crest. Both appeared on a set of Icelandic postage stamps depicting eight of this country's fabled monsters, issued on March 19, 2009.

There is even a possibility that some of the Orient's fabled water dragons were inspired by giant oarfish sightings, albeit highly unusual ones—judging at least from the following modern-day report if true. In his book *Very Crazy, G.I.: Strange But True Stories of the Vietnam War* (2001), Vietnam combat veteran Kregg P.J. Jorgenson reported that in

1999 he learnt of a truly extraordinary encounter, allegedly made by Craig Thompson, who had served in Vietnam as a 20-year-old sergeant E-5 from Coeur d'Alene, Idaho, with Company B, 2d Battalion of the 503d Parachute Infantry Regiment, 173d Airborne Brigade.

The famous photograph of a supposed nagini, clearly a giant oarfish (public domain)

Thompson claimed that his platoon had been bathing one day in the Bong Son River, in Vietnam's Binh Dinh Province, when one of the soldiers caught sight of a large serpent-like creature swimming up the river towards them. They estimated it to be at least 30 feet long, and 1-2-feet wide, covered in glistening gold scales. Of particular note was that its large square head bore a dark red plume that stood high out of the water as it swam, and its long undulating body trailed behind. Not surprisingly, Thompson and his men shouted to each another to get out of the water as they reached for their weapons; but before they had time to do so, the creature disappeared beneath the murky waters and was seen no more.

For a long time, Thompson remained perplexed as to the zoological identity of this "golden dragon," until he learnt about the giant oarfish, which is what he now believes that he and his platoon saw—and this red-crested serpentiform species does indeed come to mind when reading his report. However, there are some notable discrepancies too, because giant oarfishes are normally silver in color, not gold; they are marine and typically mesopelagic, not freshwater and surface-dwelling; and even in those very rare instances when they are seen swimming near the surface, they do not do so with their crest standing high out of the water. So if Thompson's report is genuine, how can these anomalies be explained?

In his book, Jorgenson claimed that gold and brown versions of the giant oarfish have been found in Australia and also off Mexico, but I have not seen pictures of any. Reflected sunlight and/or river silt clinging

RITRATTO DEL MVSEO DI FERRANTE IMPERATO

Ferrante
Imperato's cabinet
of curiosities,
featuring a giant
oarfish (arrowed)
(public domain)

to its scales may conceivably have rendered the Vietnam creature golden in appearance, but in view of how closely and clearly it was seen (and by so many witnesses, not just one), both options seem rather unlikely.

Equally, I am not aware of any records of the giant oarfish turning up in rivers, nor indeed of it swimming anywhere at all with its red crest raised up above the water surface. Certain moray eels in the Far East do venture into freshwater and have golden-brown scales, but they are far smaller than the dimensions offered for the Vietnam creature, and, crucially, they do not possess the latter's high red crest, which is a diagnostic characteristic of the giant oarfish. At present, therefore, Thompson's freshwater "golden dragon" oarfish remains an unresolved enigma.

Yet if such a creature (or cryptid, as it assuredly must be) really does exist, and regardless of whether or not it actually is a giant oarfish, it may help to explain at least some of the ancient Far Eastern legends of golden freshwater dragons, and also of crested nagas with gilded scales—as regally portrayed, for instance, by various imposing statues at Bangkok's Royal Palaces in Thailand.

Incidentally, although the giant oarfish was not formally recognized by science until Ascanius's description of it in 1772, the myth of Laocoön's destruction is not the only evidence that this mysterious, little-seen, yet instantly-recognizable species had been known long before then. Direct confirmation of this comes from the fact that a preserved giant oarfish was present in the famous cabinet of curiosities displayed at Palazzo Gravina in Naples, Italy, by Ferrante Imperato, a Neapolitan apothecary. He referred to this specimen as *Spada marina* ("sea sword") in his *Dell'Historia Naturale* (1599)—a multi-volume catalogue of his cabinet of curiosities, which contains a plate depicting the latter with the giant oarfish clearly visible upon one of its walls.

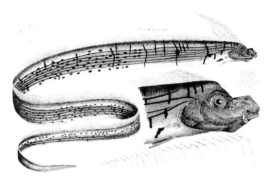

Beautiful engraving of a giant oarfish, with a close-up of its surprisingly equine head and protrusible toothless jaws (public domain)

I'll leave the final words on the giant oarfish to the late Arthur C. Clarke, one of whose characters in his classic sea monster-featuring science-fiction novel *The Deep Range* (1957) voiced the following apt description and equally telling cryptozoological sentiment:

> ...but the really spectacular one is the oarfish—*Regalecus glesne*. That's got a face like a horse, a crest of brilliant red quills like an Indian brave's headdress—and a snakelike body which may be seventy feet long. Since we know that these things exist, how do you expect us to be surprised at anything the sea can produce?

Amen to that!

Chapter 11:
WHY BLOOD-DRAINED CARCASES ARE *NOT* THE WORK OF CHUPACABRAS OR OTHER SUPPOSEDLY VAMPIRIC CRYPTIDS

The thing about myths is that people want to believe in things. I suppose that, in a perverse way, there's something comforting in that there's this vampiric monster that doesn't attack humans.

— Ben Radford interviewed by Bjorn Carey, nbcnews.com, March 22, 2011

Discoveries of supposedly blood-drained animal carcases hit the cryptozoology headlines with monotonous frequency, accompanied by the usual (and sometimes decidedly unusual) media speculation as to what diabolical entity could have been responsible for such a hideous, unnatural act.

In reality, of course, no such entity—diabolical, vampiric, or otherwise—is responsible, because it is highly unlikely that these carcases really are blood-drained (variously termed desanguinated or exsanguinated). They merely look as if they are, particularly to the pathology-untrained eye, which is a very different matter altogether.

Over the years, many culprits for such unsavory activity have been proposed—the chupacabra or goatsucker being the favorite identity if said carcases have been discovered in the New World; and various mystery carnivores, such as escapee/released big cats and even the (very) odd absconded, far-from-home thylacine, if elsewhere.

Ironically, however, the true nature of these carcases has already been investigated, uncovered, and publicly exposed for all to see and read about in a quite recent cryptozoology book that I heartily recommend to everyone—Benjamin Radford's superb *Tracking the Chupacabra: The Vampire Beast in Fact, Fiction, and Folklore* (2011). So why such carcases should continue to perplex

Reconstruction of chupacabra with victim (William Rebsamen)

other researchers and the media, as indeed they still do, thoroughly baffles me.

Here's the original, unedited version of my review of Ben's book that was subsequently published in slightly abridged form by *Fortean Times* (issue #285, March 2012):

In modern times, very few cryptids have risen from obscurity to international celebrity with such alacrity as the chupacabra or goatsucker (indeed, I can only think of one other offhand—the Mongolian death worm). Prior to the 1990s, it was a Hispanic oddity, now it is a by-word for mysterious entities of the deliciously dark and sinister kind—emblazoned as a snarling, fiery-eyed anti-hero upon t-shirts sold in every corner of the planet, and gorily eviscerating and exsanguinating its hapless victims in saliva-dripping glee as the toothy vampiric star of countless movie and video flicks viewed worldwide.

But what, precisely, is the chupacabra, and where did it come from? Indeed, does it even exist? Over the years since it first began hitting the media headlines in Latin America and then steadily onward and outward until its infamy became a global sensation, this monstrous marauder has been described in countless different ways by its supposed eyewitnesses—likened to just about everything, in fact, from a spiky-backed bipedal pseudo-kangaroo with wings and spinning hypnotic eyes to a hairless quadrupedal blue dog with mangy demeanour and long savage jaws. Its origin has attracted equally diverse, dramatic speculation too—a spontaneously-mutated freak of nature, an absconded scientific experiment gone wrong, even a decidedly inimical extraterrestrial visitor.

It was high time, therefore, that this paranormal Proteus received an in-depth, critical scientific examination, seeking both its identity and its origin; and thanks to this riveting book, that is exactly what it has received. As in his previous works, Radford has painstakingly stripped away the layers of glamour, hearsay, folklore, and media hype to reveal what he believes to be the truth behind the lurid crypto-legend, the reality at the heart of this unlikeliest of contemporary icons, and I for one consider that he has achieved his goal.

Along the way, as with all of the most thorough investigations, there have been a number of stark, surprising revelations. Not least of these, following his forensic examination of the case in question, is Radford's comprehensive dismembering of the first major, pivotal eyewitness report (which had almost single-handedly launched the chupacabra phenomenon in fully-formed state upon an unsuspecting world).

Also well worthy of mention here but without giving away the all-important details is his documentation of the long-awaited explanation for why supposed chupacabra victims' carcases are often described as being entirely drained of blood; as well as how another perplexing entity, the reptilian humanoid of Thetis Lake in British Columbia, Canada, was lately exposed as a hoax—a significant event, yet which had not previously received widespread coverage. Those pesky blue dogs sans hair reported from Texas and elsewhere in recent years and even represented in the flesh by one or two preserved corpses also receive Radford's full attention, revealing their identity to be intriguing but far less outstanding than media reports would suggest.

Radford's central thesis, however, concerns the remarkable but hitherto uncommented-upon similarities between the chupacabra and the alien star of a certain science-fiction film whose release occurred just prior to the first, crucial eyewitness report of el chupa to attract major media attention. Was the latter shaped by the former? Judging from the evidence presented here by Radford, this would certainly seem to be the case, influencing everything written about and described for the chupacabra since then.

After spending far too many years in the headlines as a bloodthirsty monster with a rapacious appetite for victims and headlines in equal measure, it looks very much as if the chupacabra has finally met its match— assassinated not with a shotgun, but instead with sterling detective work. Consequently, I feel it only fair to warn you that if you like your newly-slain goatsucker served with a generous dollop of mystery and spiced with all manner of rarefied unsubstantiated rumours, you are not going to enjoy this book. If, conversely, you prefer it plucked raw and served cold, basted only by scientific detachment and common sense, it should be a veritable feast.

The most popular representation of the chupacabra's reputed form (public domain)

So, returning to the subject of desanguinated carcasses, what precisely is their true explanation? As Ben revealed in his book's concluding, 35-page chapter, entitled "The Zoology of Chupacabras and the Science of Vampires" and comprising what I consider to be the most forensic, rigorous examination of this subject ever published within the cryptozoological literature, the answer is not even remotely preternatural, but is in fact remarkably, unexpectedly mundane. Summarizing his revelations, here are the most salient points:

1) Some such reports documented by the media are not first-hand but rather second-, third-, or even fourth-hand, and are thereby susceptible to distortion and fabrication of the "Chinese whispers" and foaflore (friend-of-a-friend lore) nature.

2) Reports that *are* first-hand originate directly from those who have discovered such carcasses, but such persons, e.g. farmers, ranchers, do not generally have medical or forensic expertise, and the carcasses themselves are very rarely examined by anyone who has. So their claims that the carcasses lack blood are not scientifically substantiated and are therefore merely unsupported personal opinion, i.e. supposition.

3) If little blood is seen on or around the carcass, a layman discoverer is likely to assume that the carcass has been desanguinated and even more likely to assume this if he should actually cut the carcass open and find little or no evidence of blood inside it. However, this apparent absence of blood is in reality no such thing. When an animal (or human) dies, rigor mortis is accompanied by livor mortis—a lesser-known process in which the carcass's blood soon begins to settle via gravity in the lower, underneath areas (which thus acquire a dark reddish-purple hue) and coagulates there, both inside vessels and in tissue surrounding vessels from which it has leaked. This only takes a few hours at most, so unless someone finding a carcass turns it over, thereby revealing the dark hue of the tissues underneath where blood has collected and coagulated (and not many people would see any reason to do so, especially with a hefty, smelly carcass like that of a dead cow or horse), the activity of livor mortis will remain hidden from view. All that will be seen is the carcass's much paler upper portion, from which blood has drained out, down into the concealed lower portions underneath.

4) Even if the carcass *is* turned over, if it has been lying on hard or rough ground the blood vessels in its undersurface tissues will have been compressed by the ground, thus restricting the settling of blood

(i.e. livor mortis) there. So this surface will appear paler (and hence more bloodless) than would otherwise have been the case.

5) If the carcass is of an animal with dark and/or very hairy skin, livor mortis-induced discoloration will not be discerned anyway, even if the carcass is turned over, unless painstakingly examined by a medical pathologist or veterinarian via a full autopsy.

6) If a carcass is cut open and little or no blood emerges, this is due merely to the fact that it has had time to become fixed in the tissues and clotted. In short, the blood is still there, but it has simply dried up and its water content evaporated.

7) To determine scientifically the extent of blood loss, or whether there has actually been any blood loss at all, from a carcass, a full-scale formal necropsy would be required, performed by a qualified medical pathologist or veterinarian. This rarely happens with animal carcasses found by farmers and ranchers on their lands, if only because of the high fee that the farmer or land-owner would be required to pay in order for such a procedure to be conducted.

8) One important indicator of significant blood loss is noticeable paleness of the internal organs, but again, unless the carcass has been professionally necropsied, this would not be readily perceived.

9) Crucially, in cases where supposedly desanguinated carcasses *have* been examined by medical or forensic experts, they have not observed anything that they have considered to be anomalous—everything present has been in accord with their professional experience of the appearance of corpses, externally and internally. Perhaps the best-known example of this is the work of David Morales, a Puerto Rican veterinarian with the Department of Agriculture. Despite having examined 300 supposedly desanguinated animal carcasses that had been blamed upon the chupacabra in Puerto Rico, he found no evidence whatsoever to support such a claim. On the contrary, he found lots of blood inside the carcasses, with no sign of vampirism, but plenty of signs that the animals had been attacked and killed by normal, mundane predators, such as dogs, monkeys, and birds.

Another interpretation of the chupacabra, highlighting its ferocious-looking dentition (Tim Morris)

In short, the blood-draining, vampiric activity of the chupacabra and other predators is a fallacy, engendered by a lack of specialized forensic, medical knowledge by those discovering and observing the carcasses, as well as by exaggerated, inflamed media accounts.

So the next time that you read about a mysteriously desanguinated animal carcass, remember the above checklist, and if the carcass hasn't been subjected to a thorough autopsy by a qualified pathologist or veterinarian, the chances are that it will be its bloodless state that is non-existent, not its blood.

For full details regarding the alleged desanguination of animal carcasses, please do read Ben's fascinating book, *Tracking the Chupacabra*—a compelling, eye-opening, and indispensable foray into the chupacabra's origin, as well as the myths, and the many fallacies surrounding this modern-day cryptozoological megastar.

Finally, I was nothing if not startled recently to discover the following chupacabra-like monster crouching near the bottom of f 97v (folio 97 verso) in an illuminated Book of Hours devotional manuscript entitled *Hours of Joanna the Mad*. This manuscript had originally been owned by Joanna of Castile, the (controversially) mentally-ill consort of Philip the Handsome, king of Castile, and had been produced for her in the city of Bruges (in what is now Belgium) some time between 1486 and 1506—almost 500 years before the chupacabra first hit the news headlines worldwide! An oft-quoted line from the Book of Ecclesiastes in the Bible's Old Testament is "There is nothing new under the sun." How true this is, even, it would seem, in cryptozoology!

The chupacabra-like monster from *Hours of Joanna the Mad* (public domain)

Chapter 12:

HORNED RODENTS, DEVIL'S CORKSCREWS, AND TERRIBLE SNAILS—REAL-LIFE CRYPTO-PALAEONTOLOGICAL DETECTIVE STORIES

The writer has recently described part of the skull of a Mylagaulus from the Colorado Loup Fork beds, found in 1898. A nearly complete skull, with one ramus of the lower jaw, found by Mr. Brown of the Expedition of 1901, indicates a new genus of this family, distinguished by the unique character (for a rodent) of a pair of large connate processes on the nasals resembling the horn-cores of some Ungulata, and giving the skull a profile absurdly like that of a miniature rhinoceros.

— William D. Matthew,
"A Horned Rodent From The Colorado Miocene,"
*Bulletin of the American Museum
of Natural History* (1902)

Horned rodents, devil's corkscrews, and terrible snails may not seem to have a lot in common, but in reality these three ostensibly separate strands are intricately intertwined within a singularly unusual and fascinating chapter in the history of monstrous zoological discoveries, as now revealed.

It all began in 1891, when geologist Erwin H. Barbour from the University of Nebraska was shown some extraordinary formations by local rancher Charles E. Holmes in the Badlands of northwestern Nebraska, USA. Barbour and Holmes colloquially dubbed them "devil's corkscrews," because they did indeed resemble gigantic subterranean screws, each one penetrating many feet below the earth's surface, and constituting an elongated spiral of hardened earth.

Barbour proposed that these were the fossilized remains of giant freshwater sponges, his theory having been influenced by the belief current at that time that the deposits in which they occurred, and which dated to the Miocene epoch approximately 20 million years ago, were the remains of a huge

Daimonelix, an illustration from 1892 (public domain)

freshwater lake.

Moreover, recalling the informal "devil's corkscrew" nickname that he and Holmes had coined for them, in a short paper published by the journal *Science* in 1892 Barbour gave to these perplexing structures the formal scientific name *Daimonelix* ("devil's screw"), sometimes spelled *Daimonhelix* or *Daemonelix* in later works. Not everyone, however, was convinced by his theory that they were prehistoric sponges.

A number of authorities favored the possibility that they were artifacts, each one having been created by the intertwining of roots from some form of prehistoric plant that had subsequently rotted away (or even by pairs of prehistoric plants, one coiling tightly around the other), with the spiral-shaped space that they had left behind becoming filled with mud, ultimately yielding one of these remarkable giant underground "screws." And once subsequent research had shown that the deposits containing them were not the remains of a lake at all but were associated with semi-arid grassland instead, even Barbour quietly abandoned his freshwater sponge proposal in favor of the plant theory.

However, the name *Daimonelix* remained valid, because although scientific genera and species names are generally given only to organisms (modern-day or fossil), a notable exception to this nomenclatural rule concerns ichnofossils or trace fossils. These are fossils not of organisms themselves but of the traces left behind by them, such as footprints, burrows, coprolites, feeding marks, plant root cavities, etc, and they too receive scientific genera and (sometimes) species names.

A third theory concerning the nature of the devil's corkscrews was put forward by Theodor Fuchs and Edward Drinker Cope, who independently suggested in 1893 that they were the fossilized burrows of a

Daimonelix, fossil rodent burrow, Sioux County, Nebraska, early Miocene, close-up (public domain)

Miocene rodent. This notion attracted appreciable interest—but if true, what kind of rodent could have been responsible? One candidate favored in various popular-format publications for quite some time during the 20th century was a creature no less extraordinary than the corkscrews themselves.

During 1902, William D. Matthew published a paper in the *Bulletin of the American Museum of Natural History* formally describing a new species of fossil rodent from Colorado that dated back to the Miocene, but which was so different from all previously recorded species that it also required the creation of a new genus. Based upon a skull found in 1898, he named this novel creature *Ceratogaulus rhinocerus* [sic]—a very apt name, because, unique among all rodents at that time, it sported a pair of short but very distinctive vertically-oriented horns, sited laterally upon the dorsal surface of its nasal bones' posterior section.

In later years, three additional horned species were discovered and named—*C. anecdotus*, *C. hatcheri*, and *C. minor*. Some of these were initially housed in a separate genus, *Epigaulus* (created in 1907), and *C. minor* has been reassigned by some workers to the related genus *Mylagaulus*, but the current consensus is that all four belong to *Ceratogaulus*. In addition, a fifth horned species, but which unequivocally belongs to the genus *Mylagaulus* rather than *Ceratogaulus*, was scientifically described as recently as 2012, by Nick Czaplewski, vertebrate palaeontology curator at the Sam Noble Oklahoma Museum of Natural History. Formally named *Mylagaulus cornusaulax*, it lived in western Oklahoma during the Miocene (interestingly, its type specimen had been overlooked in a storage unit at the museum for 50 years before being belatedly discovered there in 2010 and recognized to be something new and special). Four other *Mylagaulus* species (not counting *C. minor* if classed as belonging to this genus) are also known, but none of these was horned.

Ceratogaulus [aka *Epigaulus*] *hatcheri*, illustration from 1913 (public domain)

Known technically and collectively as mylagaulids, the horned rodents and several closely related genera of non-horned species constitute an entirely extinct taxonomic family, existing from the Miocene to the Pliocene and (in the case of the horned species) unique to North America, but belonging to the squirrel lineage of rodents (Sciuromorpha). Moreover, examination of complete and near-complete skeletal remains has revealed that they superficially resembled marmots and other ground squirrels too, both in size (measuring roughly two feet long) and in overall appearance—except of course for the five horned species' nasal horns, which make them the smallest horned mammals known to science. The horned species are sometimes colloquially referred to as horned gophers, but this is a misnomer, because gophers are only very distantly related to them. "Horned marmot" would be a much more appropriate name.

Suggestions that the devil's corkscrews could be the fossilized remains of burrows excavated by these rodents, utilizing their horns, attracted interest and remained in contention as the solution to this longstanding mystery until as recently as the 1970s (my *How and Why Wonder Book of Prehistoric Mammals* was still supporting it back in 1964 when bought for me as a young child by my mother). However, a study by Samantha S.B. Hopkins published in 2005, focusing upon their horns' precise conformation and speculating upon what this indicated relative to their possible functions, revealed that such an idea was inherently and fatally flawed. Both the position and the shape of the horns are inconsistent with their being efficient digging tools.

By being located on the posterior rather than the anterior section of the nasal bones, the horns could not be used for digging through earth without the animal's muzzle constantly getting in the way, severely impeding the efficiency of this activity. Moreover, in later species the horns were positioned even further back than in the earlier ones, so it is evident that these rodents' evolutionary development became increasingly contrary to their horns being used as digging tools. The horns' very broad, thick shape also argued persuasively against their effectiveness as digging tools (it is nowadays believed that they served as defensive weapons instead). And so too did the telling fact that no remains of horned rodents discovered in direct association with devil's corkscrews has ever been documented.

But if the horned rodents were not responsible for these structures, then what was? As far back as 1905, Olaf A. Peterson from the Carnegie Museum in Pittsburgh, Pennsylvania, USA, had revealed that some of them contained fossilized bones from *Palaeocastor fossor* and *P. magnus*—two prehistoric species of small terrestrial beaver. These had existed in Nebraska and elsewhere in North America's Great Plains region during the late Oligocene and Miocene epochs. However, it was not until 1977 that their responsibility for creating the devil's corkscrews was confirmed, via a paper published in the scientific journal *Palaeogeography, Palaeoclimatology, Palaeoecology*, and authored by Larry D. Martin and D.K. Bennett.

In it, the authors disclosed that these enigmatic underground spirals were in fact the helical shaft sections of *Palaeocastor* burrows, each complete burrow consisting of a single entrance mound, a long spiraled shaft, and a lower living chamber. These burrows also possessed interconnecting side-passages, and the authors' paper revealed that very extensive subterranean *Palaeocastor* colonies had existed (Martin had discovered one that contained over 200 separate burrows). Indeed, they were comparable in size and network complexity to the labyrinthine underground "towns" or "cities" produced by those modern-day North American ground squirrels known as prairie dogs (*Cynomys* spp.).

In addition, Martin's research at the University of Kansas had uncovered that the beavers excavated these screw-shaped burrow shafts with their incisor teeth, not with their claws (as various previous proponents of a rodent origin for such structures had wrongly assumed). For instead of finding narrow claw marks on the burrow walls, which is what he had expected, Martin instead discovered numerous broad grooves—which he was able to duplicate exactly by scraping the incisors of fossil *Palaeocastor* skulls into wet sand. The very regular spirals of their burrows' shafts (i.e. the devil's corkscrews) had been constructed by the beavers via a continuous series of either left-handed or right-handed incisor strokes.

And as final proof that *Palaeocastor* was indeed the engineer of the devil's corkscrews, the wider chambers immediately

Palaeocastor fossil remains inside burrow's living chamber (public domain)

below these spiraled shafts were sometimes found to contain perfectly preserved fossil skeletons of adult beavers and beaver cubs, thereby verifying that they were the burrows' living quarters for these beavers.

After almost a century, the mystery of North America's devil's corkscrews was a mystery no more; but across the Atlantic in England, an equally spectacular edifice of spiraled structure has continued to baffle scientists. Its name? *Dinocochlea*—"the terrible snail."

In 1921, during the construction of a new arterial road near Hastings in the Wealden area of Sussex, an enormous spiral-shaped object was uncovered and excavated from early Cretaceous clay after having been spotted by site engineer H.L. Tucker. Outwardly it resembled the spiraled shell of certain marine gastropod molluscs, in particular those of the genus *Turritella*, which is represented by numerous living and fossil species.

Accordingly, when it was formally described during 1922 by London's Natural History Museum mollusc specialist Bernard B. Woodward in the *Geological Magazine*, he named it *Dinocochlea ingens* and did indeed categorize it as a fossil gastropod, albeit one of immense proportions.

Measuring more than six feet in length, it was far bigger than any other gastropod species known then, or now. However, this identification incited much controversy.

Dinocochlea ingens, 1922 newspaper image (public domain)

For whereas spiraled gastropod shells normally bear ridges and possess coils that taper to a point, *Dinocochlea* did not, and there were no shell traces preserved with it either. Its freakishly large size was also difficult to reconcile with a gastropod identity.

Recalling the devil's corkscrews of North America, was it possible, therefore, that *Dinocochlea* was actually the fossilized burrow of some still-undiscovered species of prehistoric rodent? Alternatively, bearing in mind that it was uncovered near to a quarry famous for the quantity of *Iguanodon* and other giant reptilian fossils discovered there, could it be a dinosaur coprolite (fossilized fecal deposit)? Once again, however, its gargantuan size (even for a coprolite of dinosaur origin!) and also its spiraled shape's very precise, regular form argued against this, as did the fact that there was no partially-digested organic material associated with it, which is normally the case with preserved coprolites. So what could this very curious, anomalous object be?

In June 2011, palaeontologists Paul Taylor and Consuelo Sendino from London's Natural History Museum (where *Dinocochlea* had been deposited following its discovery) presented a new and very plausible explanation. In a *Proceedings of the Geologists' Association* paper, they proposed that it had indeed originated as a corkscrew-shaped burrow, but a horizontal one rather than the vertically-oriented devil's corkscrews, and had not been created by any rodent but instead by a fossil species of capitellid polychaete worm known as a threadworm. Yet as these were less than an inch in diameter, how could so tiny a creature have produced such a monstrously huge trace fossil as *Dinocochlea*?

Having examined cross-section specimens of it, which revealed that they were filled with concentric bands of sediment resembling the growth rings of tree trunks, Taylor and Sendino suggested that although initially very small, this worm burrow had acted as a nucleus for concretion growth (which is characterized by the presence of such rings or bands internally). That is, the space originally created by the burrow would induce the movement into it of surrounding mineral cements, which would themselves then leave behind a space that would in turn induce the movement into it of more surrounding cements, and so on. Consequently, if conditions for its preservation were just right, what began as a tiny thin worm burrow would ultimately become enormously enlarged, yielding the very dramatic pseudo-gastropod, mega-burrow trace fossil that we know today as *Dinocochlea*.

From horned marmots and burrow-digging beavers to devil's corkscrews and terrible snails-that-weren't, it is evident that however distant our planet's past may be, it still possesses the power to perplex, surprise, inform, and fascinate us in a myriad of different ways.

Life-sized *Dinocochlea* model and Dr Paul Taylor from London's NHM (public domain)

Chapter 13:
BOTHERSOME BEITHIRS, THE LOCH NESS MONSTER, AND OTHER FRESHWATER MYSTERY EELS?

Many things have been said about Swanson's river. There were tales of a giant pike in the northern pond whose eyes were blind and a red roach served as its guide. There was a gigantic eel with large horns on its head.

— Fritiof Nilsson Piraten, Bombi Bitt and Me

Large, dark shapes began to appear, attracted by the bread that Beryl fed her ducks.

"I don't know where they'd come from. I guess they'd always been there." She looked into the pool. "They just keep getting bigger and bigger..."...She tied a piece of steak to a string, and we watched her wade out into the shallow pool in her gumboots. As she waved the steak in the current, I saw a few large heads emerge from the watercress. Giving in to a natural reflex, I took a step back...

As Beryl lifted the steak on the line out of the water, a huge eel, about as big around as the calf of her leg, lifted its head out, dancing to-and-fro to keep its body up, not unlike a cobra...

When she lowered the meat into the water, five or six big eels, their heads five to eight inches across the back, vied for a piece. They grabbed on, making loud sucking sounds as they tried to get an advantage on the steak, rolling their bodies to tear pieces off.

Stella had taken off her flip-flops and was walking barefoot across the grass. She spread one of the cans of dog food near the edge of a pool. With a stick, she pushed some of the meat chunks toward the water. A single big eel came to the rim of the concrete ledge to investigate. It sniffed a few times, then tilted its head and body, propelled itself over the ledge onto the grass, and began taking pieces of the dog food in the side of its mouth. A few smaller eels followed, and soon the grass was wet from eel slime. They had no trouble coming completely out of the water to take the food.

— James Prosek, "Maori Eels,"
Orion Magazine (July/August 2010)

The species featuring in the previous page's quote is New Zealand's longfin eel *Anguilla dieffenbachii*, whose adult females are much longer than the males and are known to attain a very impressive total length of up to five feet. However, traditional Maori legends tell of even bigger, truly monstrous specimens—and so too do the annals of cryptozoology. Indeed, a number of serpentiform freshwater cryptids reported from several very different localities around the world might conceivably be unusually—if not exceptionally—large eels. This thought-provoking possibility is explored here by way of the following selection of eye-opening examples.

NEVER BOTH A BEITHIR

The Loch Ness monster (LNM) may well be Scotland's best-known freshwater mystery beast, but it is not this country's only one. Far less familiar yet no less intriguing in its own way is the beithir. In 1994, a correspondent to the English magazine *Athene* published two fascinating articles containing various modern-day beithir sightings. During early 1975, he encountered a fisherman near Inverness who claimed that he and four others once sighted a beithir lying coiled in shallow water close to the edge of a deep gorge upstream of the Falls of Kilmorack. When it realized that it had been observed, however, it thrashed wildly about before finally swimming up the gorge near Beaufort Castle and disappearing. The fishermen estimated its length at around 10 feet.

Four months later, the same *Athene* correspondent learnt of another sighting, this time offshore of Eilean Aigas, an island in the River Beauly, Highland. He was also informed by a keeper at Strathmore that during the 1930s his wife's parents had seen beithirs moving overland at Loch a' Mhuillidh, near

The European eel (public domain)

Glen Strathfarrar and the mountain of Sgurr na Lapaich. After discussing these reports with various zoological colleagues, he considered that the beithir was probably an extra-large variety of eel—fishes that are well known for their ability to leave the water and move overland to forage when circumstances necessitate, and even to sustain themselves out of water for protracted periods.

Indeed, the *Athene* correspondent was informed by a Devon farmer that during the extremely harsh winter of 1947, his mother had been badly frightened to discover a number of eels alive and well in the farm's hayloft, where they had evidently been sheltering since the freezing over of the nearby river some time earlier. The rest of the family came to see this wonder, including the farmer himself (then still a boy), and his father confirmed that they were indeed eels, and not snakes (as his mother had initially assumed).

IS NESSIE A EUNUCH EEL?

The LNM (always assuming that it actually exists, of course!) has been labeled as many things by many people—a surviving plesiosaur, an unknown species of long-necked seal, and a wayward sturgeon being among the most popular identities proffered over the years. However, some eyewitnesses and zoological authorities—notably the late Maurice Burton—have favored a giant eel, possibly up to 30 feet long.

Under normal circumstances, the common or European eel *Anguilla anguilla* does not exceed five feet, and even the conger eel *Conger conger* (one of the world's largest eel species, rivaled only by certain moray eels) rarely exceeds 10 feet. However, ichthyological researchers have revealed that growth in eels is more rapid in confined bodies of water (such as a loch), in water that is not subjected to seasonal temperature changes (a condition met with in the deeper portions of a deep lake, like Loch Ness), and is not uniform (some specimens grow much faster than others belonging to the same species).

Collectively, therefore, these factors support the possibility that abnormally large eels do indeed exist in Loch Ness. Moreover, divers here have claimed sightings of such fishes. Also of significance is the fact that eels will sometimes swim on their side at or near the water surface, yielding the familiar humped profile described by Nessie eyewitnesses. And an 18-30-foot-long eel could certainly produce the sizeable wakes and other water disturbances often reported for this most famous—and infamous—of all aquatic monsters.

Consequently, I would not be at all surprised if the presence of extra-large eels in Loch Ness is conclusively demonstrated one day.

Delightful artistic impression of Nessie as a serpentiform aquatic monster (Richard Pullen)

However, I cannot reconcile any kind of eel with the oft-reported vertical head-and-neck (aka "periscope") category of LNM sightings, nor with the land LNM sightings that have described a clearly visible four-limbed, long-necked, long-tailed animal.

Yet regardless of what creature these latter observations feature (assuming once again their validity), there is no reason why Loch Ness should not contain some extra-large eels too. After all, any loch that can boast a volume of roughly 1.8 cubic miles must surely have sufficient room for more than one type of monster!

In recent years, the giant eel identity for Nessie has been modified by some cryptozoological researchers to yield a creature as remarkable in itself as any bona fide monster—namely, a giant eunuch eel. It has been suggested that Nessie may be a gigantic, sterile or eunuch specimen of the common European eel—one that did not swim out to sea and spawn but instead stayed in the loch, grew exceptionally long (25-30 feet), lived to a much greater age than normal, and was rendered sterile by some currently-undetermined factor present in this and other deep, cold, northern lakes.

This is undeniably a fascinating, thought-provoking theory, but no such specimen matching this hypothesized creature has ever been brought to scientific attention. Moreover, Scott McNaught, professor of Lake Biology at Central Michigan University, has stated that even if eunuch eels did arise, they would tend to grow thicker rather than longer. Nevertheless, giant eels remain a distinct possibility in relation to some of the world's more serpentiform lake monsters on record.

THE GAS STREET MONSTER OF BIRMINGHAM, ENGLAND

In England, Birmingham is a major industrialized city at the heart of the West Midlands region where I live. Indeed, it is England's second largest city by area (only the capital, London, is bigger), and is famous for its extensive canal system—so extensive, in fact, that it has often been compared to that of Venice.

In 1997, these waterways were said to be home to a gargantuan eel, variously dubbed the Gas Street Monster (after the Gas Street canal basin where it had allegedly been sighted on several occasions) and the Brumbeast (Brum being a local nickname for Birmingham). It was described by one angler eyewitness as being black in color, "with little beady eyes," and up to 20 feet long, but nothing confirming this grandiose estimate ever materialized on the end of any of the local fishermen's lines. Reports of the monster eventually fizzled out, suggesting perhaps that Gas Street's mega-eel had moved on to canals new, or had never existed to begin with. Eels of a more traditional, modest size, conversely, measuring up to three feet or so in length, are often seen in Birmingham and other Midland canals.

Interestingly, fellow cryptozoologist Nick Redfern (who was also a fellow Midlander, living just a few miles away from me until he emigrated to Texas) has revealed that stories of a giant eel in Birmingham's canals did not begin with the Gas Street Monster in the 1990s, noting that such a creature had been reported in these waterways at least a decade earlier. On April 18, 2008, Nick posted an article entitled "The Great Eel of Birmingham" on his website *There's Something in the Woods...* that contained the following thought-provoking excerpt:

> Back in the 1980s, as I well recall, rumors began to circulate among local Forteans of sightings of a huge eel seen lurking in the dark waters of the winding canals that both surround and cut through Birmingham.
>
> One particularly memorable account originated with a lorry-driver who recalled such a sighting somewhere in Birmingham in the latter part of the 1980s; and that "shook the staff rigid" at a plumber's merchants that overlooked the stretch of canal in question.
>
> In this case, the animal was described as being dark brown in color and was said to be no less than an astonishing fifteen feet in length. Supposedly, it had been briefly seen by a fork-lift driver, who had sat, not surprisingly mesmerized, watching it "circling" one particular area of the canal frequented by a large number of semi-tame ducks that the staff at the plumber's merchant would regularly feed with bread during their daily lunch-hour.
>
> Several other such stories of a distinctly similar nature caught my attention during that long-gone era; and I'm actually in little doubt that something monstrous had indeed made the canals of

Birmingham its home—albeit for a brief period, and presumably before moving on.

Or maybe it didn't move on, but instead simply stayed there, remaining in stealthy seclusion until hitting the headlines again in 1997? And who knows, as European eels can exhibit very long life-spans, especially when in confined conditions, perhaps it is still there today...

IS ELVIS ALIVE, WELL, AND LIVING IN HASLAR LAKE?

And perhaps Elvis is still hiding out in Haslar Lake, near Gosport in Hampshire, southern England. Elvis was the apt name given by locals to a supposed "king" conger eel of gigantic proportions claimed by them to be lurking there in 1987. Eyewitnesses alleged that it was at least 12 feet long (i.e. at least two feet more than the currently-confirmed maximum length for congers), grey in color, and equipped with a very plentiful supply of teeth, and that it may have reached there from the open sea, possibly even by way of some over-land travel.

Elvis first hit the headlines when it supposedly attacked 19-year-old Ralph Marshall on August 9, 1987, after he and two friends had dived off a dinghy in the middle of the lake. Ralph needed stitches at the nearby Royal Naval Hospital for wounds to his left foot where Elvis reputedly bit him. Not long afterwards, it also allegedly chomped a chunk out of a youngster's swimming flipper while a group of schoolchildren were swimming at the lake's edge. One of them, 16-year-old James Walker, described Elvis as: "This long, shiny thing under the water. It looked like a huge iron bar but it was moving." Worth pointing out, however, is that once a mature adult conger has attained its full length, its teeth drop out. So if Elvis were a mature adult (as certainly indicated by its huge size if accurately estimated), it would be toothless, and therefore unable to bite.

In a bid to verify the reality or otherwise of Elvis, on September 16 several feet of water were drained out of Haslar Lake and into the English Channel when Maurice Ponting and Bob Taylor from Gosport Borough's Sewage Division opened the weir gates. However, the only anguilline entity discovered in this then much shallower lake was a relatively diminutive two-foot-long eel duly dubbed Tiny Tim. Undaunted, the following day Stuart Beavan and Lee Marshfield from Southsea's Sealife Centre waded into Haslar, wielding giant nets in the hope of capturing the elusive Elvis alive and exhibiting it at their centre. By nightfall, however, they had conceded defeat, with no sightings let alone a successful capture of Elvis having occurred. Since then, Elvis has faded from the headlines and into local folklore.

MONSTER EELS IN THE MASCARENES

The concept of giant freshwater eels is by no means limited to Britain. For example: a number of deep pools in the Mascarene island of Réunion, near Mauritius, in the Indian Ocean, are supposedly inhabited by gigantic landlocked eels.

In a letter to *The Field* magazine, published on February 10, 1934, Courtenay Bennett recalled that during the 1890s when Consul at Réunion he saw a dead giant eel that had been caught in one such pool, the Mare à Poule d'Eaux, which is said to be very deep in places. The eel was so immense that "steaks as thick as a man's thighs were cut" from its flesh.

According to native testimony, moreover, during the heavy winter rains the giant eels could apparently be seen circling along the sides of this pool, searching for a way out. Being so exposed, however, they were prime targets for local hunters, who would catch them using a harpoon and a rope hitched round a tree. Their flesh would then be sold for food in a neighboring village.

Marbled eel (public domain)

The best-known species of eel known to inhabit freshwater on Réunion is the marbled eel *Anguilla marmorata*, the world's most widely-distributed species of anguillid (true eel), recorded extensively in the tropical Indo-West Pacific and adjacent freshwater habitats, ranging from southeastern Africa to the Society Islands (French Polynesia) north to southern Japan. Named after its distinctive marbled or mottled patterning, it is known to attain a maximum length of 6.5 feet, but if the giant Mascarene eels documented here belong to this species, then as long as their size has not been exaggerated it may exceed this confirmed length.

BEWARE OF THE BOILING POT MONSTER!

One of the world's most spectacular natural wonders is Victoria Falls, situated on the Zambia-Zimbabwe border and created by the Zambezi River when it plunges over a sheer precipice to a maximum drop of 355 feet, down into a chasm known as the First Gorge, which varies from 80 feet to 240 feet wide. The only outlet from this chasm is a narrow channel through which the entire Zambezi River duly forces itself, at the end of which is a deep pool known as the Boiling Pot, which has a diameter of approximately 500 feet, and exhibits very heavy, boiling turbulence at high water. It is here where reports of a monstrous eel-like beast have been spasmodically reported.

Perhaps the most detailed account of the Boiling Pot monster is that of W.L. Speight, writing in a little-known article on African mystery beasts that was published in 1940 by the periodical *Empire Review*. As far as I am aware, however, Speight's account has never appeared in any cryptozoological book, so I have pleasure in reproducing it here:

> At the Victoria Falls, occasional rumours are heard of a strange monster supposed to live in the Boiling Pot. This creature is like a gigantic eel, pale in colour, with a glossy skin. Those who have seen it say it moves easily through the fiercely convulsed pool and then disappears. This monster seems first to have been noted by two Englishmen holidaying at the Falls. They told their story over the dinner table, and were laughed at. In spite of their amusement, many guests watched the Boiling Pot next day, but saw no sign of the monster. The idea was treated as a joke and, in the course of time, forgotten.
>
> Many years later the monster was seen again. Two women tourists were standing above the Falls at noon when a creature with a long head and a great length of body suddenly appeared in the Boiling Pot. It moved about as though searching for something and then disappeared. The startled ladies waited a long time, but the monster did not reappear. This story was corroborated by the man who took tourists on launch trips. Both incidents occurred several years ago, and since then the monster does not appear to have been seen. Some think it was the creation of an excited brain,

and although it is difficult to believe that such a monster lives at the Victoria Falls, the fact that more than one person and at different times has reported seeing it seems to give substance to the story. Even to the most prosaic the Victoria Falls suggests the unbelievable. The Dutch explorers who saw the Falls long before Livingstone used to hint vaguely at unknown terrors. Is it possible that they and the natives who knew "the smoke that thunders" in the early days had seen some gigantic and mysterious creature in the deep waters into which the Falls tumble? If the monster does not reveal itself again we will never know.

If not an eel, perhaps it was a very large water snake, but without further details, particularly morphological ones, nothing conclusive regarding the Boiling Pot monster's identity can be proposed.

EXTRA-LARGE EELS IN JAPAN?

Several of Japan's biggest lakes are associated with accounts of freshwater eels reputedly much larger than typical specimens on record from these localities. A concise coverage of such creatures appeared in a detailed article concerning Japanese giant mystery fishes that was written by Japan-based cryptozoology enthusiast Brent Swancer and posted on April 30, 2014, at the *Mysterious Universe* website, and reads as follows:

> Various locations in Japan have had reports of huge eels far larger than any known native species.
>
> Workers doing construction on a floodgate on the Edo river reported coming across enormous eels measuring 2 meters (6.6 feet) long. According to the account, four of the eels were spotted and some of the workers even attempted to capture one, as the eels appeared to be rather lethargic and slow moving. They were unsuccessful as they did not have the equipment to properly catch one. Upon returning to the scene later on with the tools they needed, they found that the mysterious giant eels were nowhere to be seen.
>
> Another account comes from Lake Biwa, which is in Shiga Prefecture, and is the largest freshwater lake in Japan. In the 1980s, there were several reports of giant eels inhabiting the lake.
>
> One such sighting was made by a large group of people aboard one of the lake's many pleasure boats. Startled ferry passengers reported seeing several very large eels swimming at the surface far from shore. The eels were described as being around 3 meters (10 feet) long, and a silvery blue color. The eels appeared to be leisurely gliding along beside the boat and were observed for around 15

minutes before moving off out of sight.

A fisherman on the same lake reported actually hooking and reeling in an eel that was reported to be around 8 feet in length. In this case, the eel was kept and eaten. Another fisherman on the lake reported seeing a similarly sized eel rooting through mud in shallow water near the shore.

Interestingly, the giant blue eels of Lake Biwa readily recall comparably described mystery beasts from India's Ganges River, as reported by several early chroniclers, and which I document in Chapter 15 of this book.

NO QUALMS IN GUAM CONCERNING GIANT EELS

Guam is the largest island of Micronesia in the western Pacific Ocean and its political status is that of an organized, unincorporated territory of the USA. Cryptozoologically, its claim to fame may well be as a refuge to freshwater eels of truly monstrous size—judging at least from the following communication posted publicly to my *ShukerNature* blog on April 14, 2015, by a reader with the user name Justathought.

Here is what he wrote:

> I was stationed in Guam during the late 80's/early 90's, at the Naval Magazine there, while enlisted in the Marine Corps. The Naval Magazine there sits in the most heavily-forested part of the island, and it is a protected game reserve, with the only "lake" on the island, Fena Lake (reservoir). I will swear that I have seen many eels exceeding 8 feet in length in the streams and small rivers that run through this area, as well as having seen quite a few eels of such length sliding down the spillway into the reservoir. If you could find other Marines who patrolled the jungle there, some will have seen such eels. Maybe some local poachers can also verify having seen eels of this length in this particular area, which is/was off-limits to locals, many of whom regularly came in over the hills to hunt deer and wild pig, set their shrimp and fish traps in the streams, etc.
>
> The "lake," Fena Reservoir, was rumored to be "bottomless," and it does drop straight off from the edges, deeper than a full spool of 6 lb. test line will take a weight. Some Navy divers went diving in that lake on their off day one day, and came out claiming that there were "eels down there big enough to swallow a man whole." I don't know about that, but it is a creepy place, day or night, and the Tilapia and bass that we pulled out of there were just enormous.

There is at least one other possible cryptozoological animal in the jungle on that reserve. We called it "Bagget's Monster," after the Marine who first claimed to see it. Japanese soldiers lived in the jungle on that U.S. base until at least 1978! That's how thick the jungle is on that reserve, and it is just a fascinating, mysterious place.

Head to Guam, get a pass and some escorts from the Navy, and you will see these 8' plus eels without a doubt, and if you are brave enough to don some scuba gear and check out Fena lake, you just might see one of those giant eels, that is rumored to feed on the water buffalo, deer, and pigs that drink from its waters.

This account has never before been published in any cryptozoology book and certainly makes fascinating, thought-provoking reading. Moreover, in his book *Guam Past and Present* (1964), Charles Beardsley provided a degree of precedence by noting: "...one group of early explorers caught sizeable, edible fresh-water eels in Guam's streams."

If the size of Guam's giant freshwater eels reported by Justathought has been accurately gauged, this means that they are among the world's largest (and perhaps even *the* largest, judging from the U.S. Navy divers' claims). Moreover, as the mottled eel *A. marmorata* is known to exist here, they may belong to this species—which is also implicated as the identity of the mystery giant freshwater eels in Réunion. In any event, they are definitely worthy of further investigation—assuming that anyone out there is brave enough to venture into their watery, bottomless domain!

GIANT EELS IN OHIO?

Although giant eels are a popular identity for water monsters (of both freshwater and marine abode), the size of eels is notoriously difficult to gauge accurately in the wild due to their sinuous movements and usual lack of background scale for precise length estimation. This means that eyewitness reports of giant specimens are normally difficult to take seriously—which is why the following account is so significant.

On February 3, 2015, American Facebook friend Chris R. Richards, from Covington in Washington State, posted on the page of the Facebook group Cryptozoology the following hitherto-unpublished report of a huge freshwater eel that he and his father had witnessed during the 1990s:

I believe whole heartily in giant eels. I saw one as long as my canoe back in the later nineties. They could result in sea monster claims. Hocking River Ohio. Directly off the side of the canoe in clear water near upper part of river. At first thought it was a tree with algae in water, then saw the head and realized the "algae" was actually a frill. The animal was thicker than my arm. The head was at the front of the 15ft Coleman canoe and the tail end trailed behind my back seat. At the time this was amazing to both my father and I. Only later did I come to fully appreciate how amazing this sighting was. I got to see it the longest as we slowly passed it and I was in the back of the boat. [The eel was] 12 to 15 ft.

The frill was presumably the eel's long, low dorsal fin, which runs along almost the entire length of the body in freshwater anguillids. What makes this report so exciting is that there is an unambiguous scale present in it—the known length of the canoe, alongside which the eel was aligned, thereby making its total length very easy to ascertain.

The only such species recorded from Ohio is the common American eel *Anguilla rostrata*, which officially grows up to four feet long. Consequently, judging from the scale provided by the canoe, the eel seen by Chris and his father was 3-4 times longer than this species' official maximum size.

Assuming their report to be genuine (and I'm not aware of any reason to doubt it), there seems little option but to assume, therefore, that bona fide giant freshwater eels do indeed exist, at least in the Ohio waterways, which is a remarkable situation and clearly of notable cryptozoological interest.

American eel
(public domain)

Chapter 14:
PIGGING OUT AT CHRISTMAS—IT'S GRIM WITH THE GLOSO (AND NATTRAVNEN)

Frey himself owned a gold-bristled boar that pulled him around in a cart. Though a living creature, this Gullinbursti had been fashioned for the gods by the dwarves. Today, Gullinbursti's descendants are made of golden marzipan and sold in little cellophane bags as good luck charms at New Year's.

That takes care of the pig on top of the table; in Sweden, if you weren't careful, there might be another one underneath it. She was the Gloso, or "glowing sow," and if you knew what was good for you, you would leave three stalks of wheat standing in the field at harvest time as an offering to her. You might also set out a bowl of porridge and a few fish heads for her to consume as she passed by on Christmas Eve. You could see the Gloso coming from a long way off, for her eyes burned like coals and her bristling back shed sparks as she moved. If she found the offerings too paltry, she would stay on to haunt the dark space under the tablecloth throughout the Twelve Nights of Christmas.

— Linda Raedisch, *The Old Magic of Christmas*

No book on monsters can ignore those of a seemingly paranormal nature, so here are a couple of fascinating yet hitherto little-known examples hailing from Scandinavia.

In the USA and the UK, the animals most closely linked to Christmastime via folklore and other traditions include such familiar and generally friendly species as the robin, the reindeer, and the turkey. In Skåne and Blekinge, the two southernmost provinces of Sweden, conversely, a very different, and far more daunting, creature pervades the Season of Goodwill, and its presence is anything but good. Scarcely known outside its Scandinavian provenance, outwardly it resembles a pig, but no ordinary one, for this preternatural entity is in many ways the porcine equivalent of Britain and America's phantasmal Black Dogs, and is just as dangerous!

Most commonly referred to as the gloso or "glowing pig" (other names for it include the galoppso and the gluppso, both translating as "galloping sow"), this dire beast is grim in every sense of the word. This is because the gloso is a church grim (or kyrkogrim in Sweden), i.e. a supernatural creature derived from the spirit of an animal or person supposedly sacrificed when the foundation of a church was built, and which now protects the church and its grounds for all eternity, and cannot be killed by any normal weapon. Generally, the gloso lives either within the cemetery of the church to which it is bound, or within a mound in a field directly adjacent to that church.

Those unfortunate enough to have encountered this terrifying entity liken it in basic appearance to an enormous female domestic pig, often jet-black in color (though sometimes a glowing ghostly-white hue), but with a ridge of razor-sharp spines or bristles running down the centre of its back, a pair of huge tusks curving out from its jaws, blazing-red eyes, and the fearful yet very real ability to breathe fire. Other tangible, physical abilities attributed to the gloso, and which thereby distinguish it from insubstantial ghosts or spectres, include its predilection for devouring fresh corpses in the churchyard and for sharpening its tusks upon gravestones. It also leaves visible tracks in its wake.

Pursued by the fire-breathing gloso! (Richard Svensson)

The gloso can be encountered at any time during the year, but it is said to be at its most malign during the week linking Christmas and the New Year. And yet it is during this same week when it can also be its most beneficial— provided a certain magical rite associated with it is performed correctly. If not, however, the person performing it will not live to see in the New Year!

According to Swedish legend, on the evening of Christmas Day (and also on New Year's Eve) anyone can discover everything that will happen to them during the incoming New Year if they are brave enough to withstand an assault by the gloso. The ritual stipulates that after the sun has set, you must visit four different churches in four different parishes, walk around each church in an anti-clockwise direction, and then blow through the keyhole of each church's door. After blowing through the keyhole of the fourth church's door, if you then peer through it you will witness all of the most notable events that await you in the New Year, rushing before your eyes in a rapid stream of images like a speeded-up movie film.

But for such precious insights, you must pay a steep price—the wrath of the gloso. For it will abruptly appear and chase after you, spurting hot blasts of fire at your rear end and striving to run between your legs so that its ridge of razor-sharp bristles can rip you apart. Happily, however, if you are brave enough to attempt the feat, there is one way in which this dread beast can be pacified—by turning around and facing it, with an arm outstretched, offering it a loaf of bread. If the gloso allows you to feed it the bread, you are safe. If not...

In some variations upon this legend, the same gift of New Year foresight can be obtained by confronting the gloso at a crossroads instead. As a teenager, the maternal grandmother of Swedish artist and cryptozoologist Richard Svensson (Richard is the source of much of this chapter's gloso material) once visited a crossroads in Blekinge on New Year's Eve for the express purpose of conjuring forth the gloso—though merely to see it rather than to witness what the New Year held in store for her. (Un)fortunately, the gloso did not materialize.

The gloso is also part of a much lengthier, more complex magical ritual in which the person taking part is hoping to gain psychic talents, and this multi-stage ritual has to be performed on several different magically potent dates, including Christmas night once again. Here is how Swedish folklorist Håkan Lindh described it to me:

> The ritual was a kind of vision-quest that a person who wanted to gain psychic gifts undertook several years in a row. After a bit of fasting he went out, under absolute silence, on a night-time walk to powerful places, a graveyard, a stream running towards north, a holy well, etc, and during these walks he was given trials. One of these was Gloso, and he avoided danger by just keeping his legs together and refusing to show fear. If he did, he came to no harm and gained a bit of magic power. Next year he met something else, a dragon turned into a chicken, for example, Odin on a horse, a band of aggressive Vättar [Norse nature spirits], and so on and on. While the ceremony went on, he got visions about who would die in the different homes he passed by, who would get ill, and what he could do to cure those illnesses. He also gained material magic tools too during these walks, like bones from dead people etc.
>
> This ritual continued to be performed until c.150 years ago, and

I personally know a few who have tried it recently.

In some Swedish traditions, moreover, the gloso haunts lonely roads where murders have occurred. Håkan has mentioned to me that just a few miles north of his home village in Halland, Skåne, is one such locality (where a murder took place during a botched robbery), and that alleged sightings of the gloso have been reported there and in the woods nearby.

In a somewhat less daunting example of gloso folklore, this formidable beast passes by everyone's house on Christmas Eve, but as long as the required edible gifts of fish heads and a bowl of porridge are left outside by each home owner, it will simply consume them and pass on by. Should someone fail to leave such gifts or leave ones that the gloso deems to be insufficiently substantial, however, it will have its revenge by lurking underneath that miserly wretch's dining table throughout the entire Christmas period in wait for food.

The gloso may be a remnant of earlier Nordic legends appertaining to Gullinbursti ("Golden Hair"). Named after its golden bristles, and also known as Slidrugtanne ("Horrible Tusks"), this was the great boar that pulled the chariot of the Norse deity Frey, god of fertility and pleasure. Moreover, in Blekinge there is even a local myth neatly combining Norse tradition with Christianity, in which every year St. Thomas, armed with a mighty sword, rides a tamed gloso during the Christmas week to rid the land of fatally alluring troll-maidens and other malevolent pagan beings—especially during the evening of December 21, known as Thomas's Eve. Presumably, his saintly status affords him immunity from being torn in two by his gloso's lethal back-bristles while riding it!

Frey and Gullinbursti, by Johannes Gehrts, 1901 (public domain)

As if the gloso were not terrifying enough, Skåne and Blekinge also lay claim to a second grim that is just as frightening and ferocious—the nattravnen ("night raven") or leharven. According to Richard again, this monstrous entity resembles a huge bird-like winged beast, dark in color but lacking feathers, and sometimes portrayed as quite pterodactyl-like in overall appearance.

At night, the nattravnen flies over the territory to which, as a grim, it is bound, and if anyone should wander into its domain the nattravnen will not hesitate to devour them. But this is not the only way in which someone seeing this entity could suffer as a result.

Should he and she happen to see it in flight as it passes in

Fleeing from a pterodactylian nattravnen (Richard Svensson)

front of the moon, illuminating and revealing its skeleton through its remarkably thin skin, which across its wings is peppered with holes, the observer will be stricken with agonizing pains, falling seriously ill and vomiting blood, and sometimes even passing blood in his urine for at least a week. So merely the briefest sight of a nattravnen should be avoided at all cost—and never should anyone look through the holes in its wings' skin, because this will bring certain death to the observer.

In addition, Håkan has informed me that in olden days if a person were murdered and buried secretly in a hidden grave afterwards, a stake would be forced through his corpse to prevent his vengeful spirit from materializing as a supernatural entity. But once the stake's wood had rotted, his spirit would then be freed, becoming a nattravnen, which would swiftly take wing in search of the murderer.

In such cases as these, however, the nattravnen didn't always assume the form of a bird (or pterodactyl). Instead, it sometimes became a skeleton wrapped in a black cape, or even a human skeleton sporting a large pair of wings, which made a loud noise as it flew through the air (even though the nattravnen itself was generally invisible). If a person heard one approaching, his only recourse was to lie flat, face-down, upon the ground (where traditionally the nattravnen cannot land) and hope that this foul entity would pass by, because if it came too close to him it would inflict sickness and even death.

In light of such horrors as the gloso and nattravnen, suddenly even our own Black Dogs, owlmen, mothmen, and other zooform entities seem positively tame by comparison. So I very much hope that my readers' Yuletide celebrations will always be blessed by a notable absence of fire-breathing pigs and paranormal pterodactyls!

Chapter 15:
FROM NUNG-GUAMAS TO BUDERIM BEASTS—
A NEW COLLECTION OF EXCEPTIONALLY
STRANGE CRYPTIDS

Monsters are part of our heritage. The first sailors believed that if they didn't fall off the edge of the earth, a man-eating sea serpent would devour them alive. Reports of leviathans and mermaids continue to this day. Most of these, when examined, turn out to be mistaken identifications of known animals. Yet every once in a while a rumor turns out to be something extraordinary: a previously unknown animal.

> — Karen Ehrlich and Lee Speigel,
> "Hidden Monsters," *Omni*
> (January 1983)

I kicked my way to the shallows, stood up, and lifted from the water a creature that seemed made of turquoise and lapis lazuli, with ruby belly and topaz eyes. It was shaped more or less like a salamander, but it wasn't any species I'd seen. If it hadn't been moving, I'd have thought it was some kind of gem-encrusted statue. The colors were that fresh and pure. Fish and amphibians often have a kind of resplendence just after they've been taken from the water, but this was more than that, this was resplendence.

> — David Rains Wallace, *The Turquoise Dragon*

The brilliant blue, hitherto-undiscovered species of salamander featuring in this novel is of course fictional, but if it were indeed real it would be truly extraordinary—just like all of this chapter's bona fide cryptids. It has become something of a tradition for me to devote the final chapter in each of my compendium-style books to a varied selection of outstandingly unusual and esoteric mystery beasts, including several that have previously attracted little or no widespread attention. These chapters have always attracted especial interest from my readers. Never one to break with popular tradition, I have pleasure, therefore, in presenting here my latest collection of exceptionally strange cryptids, which includes a close encounter of my own with a very sizeable, six-legged enigma.

XENOTHRIX—A MYSTERIOUS MONKEY FROM JAMAICA

Today, some monkey species inhabit Jamaica, but none of them is native; they are all South American or African species that have eventually established themselves following the escape/release of pets or other captive specimens here during the 18th century or later.

However, there is at least one enigmatic report of monkeys existing on Jamaica prior to this time. In Sir Hans Sloane's two-volume tome, *A Voyage to the Islands Madera, Barbados, Nieves, S. Christophers and Jamaica* (the two volumes were published in 1707 and 1725 respectively), when documenting the fauna of Jamaica, he included a brief but tantalizing mention of monkeys "found wild in this island." What could these have been?

On January 17, 1919, a remarkable discovery was made that may have provided the long-awaited answer to that question. That was the day when paleontologist Harold Anthony from New York's American Museum of Natural History disinterred a mandibular (lower jaw) fragment and a femur of a monkey in the yellow limestone detritus of Long Mile Cave in Jamaica's Cockpit Country. As these were discovered not too far away from some human remains, Anthony wondered whether they were from an introduced monkey specimen (the pet of a seafarer, perhaps?).

After they were collected, these bones remained undescribed and forgotten for many years, until, in 1952, two graduate students, Karl F. Koopman and Ernest E. Williams, discovered them in a drawer at the American Museum of Natural History. And when finally examined, they surprised everyone because they combined features from several different types of New World monkey. The mandible's dental formula, for instance, differed from that of all New World monkeys except for the marmosets and tamarins, but its size was much bigger than the mandible of these latter species. Further studies emphasized similarities between the Jamaican monkey and South America's titis and douroucoulis (night monkeys). Consequently, when formally described, this anomalous species was placed within a new genus, all to itself, and was christened *Xenothrix mcgregori*.

During the 1920s, Anthony uncovered some additional material from this species, including post-cranial remains such as an os coxae (a bone from the pelvic girdle), and two tibiae. As for the femur from 1919, this was not included in the 1952 description of *Xenothrix mcgregori*, remaining unstudied until 1991. When its specific form was closely assessed, however, scientists concluded that *Xenothrix mcgregori* habitually moved and climbed in a slow quadrupedal manner, while hanging upside-down from branches (and even feeding in this inverted position)—closely analogous, therefore, to a tree sloth, and thus very different from any living species of New World monkey.

Further remains, including part of the lower face from one specimen, were unearthed by various expeditions to Jamaican caves between 1994 and 1996, and these supported the notion that *X. mcgregori* was most closely related to the titis, although one researcher believes it to be a Jamaican species of douroucouli. Moreover, the precise nature of its dentition indicated that it was primarily frugivorous (a fruit-eater), and estimates of its size have proposed that it probably weighed 4.5-9.0 pounds.

In short, *X. mcgregori* was clearly a very distinct, valid species in its own right and certainly not merely based upon specimens of introduced, non-native species. But when did it die out? A partial skull and palate of *X. mcgregori* found in Lloyd's Cave near Jackson's Bay, Jamaica, were discovered in surface debris together with remains of various domestic animals and also introduced black rats *Rattus rattus*, and just like those they were unmineralized and unencrusted. This suggested that *X. mcgregori* was still alive at the time when Western explorers such as Christopher Columbus first reached the West Indies (i.e. the late 15th century). But could it have survived even later? The

The poto engraving (top) compared with a kinkajou engraving from 1849 (bottom) (public domain)

Sloane reference in the early 18th century to monkeys in Jamaica provides one intriguing indication of this, but there is another, even more baffling piece of evidence to consider too.

During the 1860s, a strange creature referred to in the publicity literature as "a poto from the mountains of Jamaica" was exhibited in

London, but the accompanying engraving of this animal depicts a beast unlike any species known to exist in Jamaica today. Consequently, some researchers have speculated that it may have been a living specimen of *X. mcgregori*. However, to my eyes it looks nothing like any type of monkey—on the contrary, what it does look very like is a kinkajou *Potos flavus*.

However, even if it was a kinkajou, that only adds to the confusion, because this small raccoon-related carnivore is not native to Jamaica either—only to Central and South America. The same is also true of its superficially similar relatives, the olingos. Moreover, it is nothing if not worthy of note that the name "poto" is very similar to "*Potos*," the kinkajou's taxonomic genus.

And as a further twist to this already much-tangled tale, when the original femur of *X. mcgregori* that had been obtained in 1919 by Anthony was finally examined during the 1990s, it was closely likened not to that of any species of monkey, but instead to that of... the kinkajou!

Some other endemic Antillean monkeys, which were related to *X. mcgregori*, have also been described from disinterred remains. These include the Hispaniolan monkey *Antillothrix bernensis* and the Cuban monkey *Paralouatta varonai*. The latter died out during the late Pleistocene epoch (and a second *Paralouatta* species died out even earlier, during the Miocene). In addition, the Haitian monkey *Insulacebus toussaintiana* was formally described in 2011 from Late Quaternary fossils obtained from the Plain of Formon, Department du Sud, in southwestern Haiti, on Hispaniola. The Hispaniolan monkey, conversely, whose remains have been found in the Dominican Republic, may have persisted until the 16th century, its demise precipitated by the island's settlement in 1492 following Columbus's arrival there. In July 2009, Walter Pickel discovered a skull of this vanished species while diving in one of Hispaniola's underwater caves.

As to whether *X. mcgregori* survived into historical times on Jamaica, however, and as to whether it is the identity of Sloane's documented "monkeys" living wild there, these remain unanswered questions three centuries later.

JUST HOW BIG—AND DEVILISH—CAN GIANT SQUIDS BE?

According to traditional Scandinavian mythology, the kraken was a monstrous sea beast that resembled a floating island, and was sometimes mistakenly landed upon by unwary sailors—who swiftly discovered their fatal mistake when the kraken's huge tentacles rose out of the sea and coiled around the sailors, dragging these hapless human victims down to a terrible watery death beneath the icy Arctic waters. During the late 1850s, however, the kraken metamorphosed from fanciful myth into zoological reality, after the remains of an extraordinary sea creature washed up onto a beach in Jutland, Denmark. These were studied by Danish zoologist Johan Japetus Steenstrup, who ascertained that they were from an unusually large squid of a hitherto undescribed species. In 1857, he formally christened it *Architeuthis monachus*—thus adding the giant squid to the zoological annals.

19th-century engraving depicting the *Alecton*'s attempted capture of the giant squid (public domain)

One of the most famous giant squid specimens hit the headlines just a few years later, and also donated some physical evidence for its species' reality. On November 30, 1861, it was spied floating in the sea 40 leagues from Tenerife in the Canary Islands by the French gunboat *Alecton*. The gunboat's captain, Lieutenant Frédérick Bouyer, and his crew bravely attempted to capture this monster, estimated by them to measure up to 24 feet long and brick-red in color, using a slipknot noose to lasso it. Unfortunately, the noose slipped down the squid's slimy body, but it became anchored around the beast's tail fin. When they tried to haul the squid aboard, however, the noose broke off the tail fin, thus releasing the rest of the squid, which fell back into the sea. The *Alecton*'s sea-giant had escaped, but its tail fin was retained, and on December 30, 1861, it was presented at a meeting of the French Academy of Sciences.

Unlike octopuses, which have only eight arms, squids have 10 arms. Eight of these are often incorrectly termed tentacles and possess suckers; the remaining two are much longer, much more slender, broad-tipped, do not possess suckers, are correctly termed tentacles, and are used for capturing prey. Since 1857, many different species of giant squid have been described, all based upon specimens washed ashore, found dying in shallows, or even regurgitated or retrieved from the stomachs of whales. Today, however, most zoologists only recognize one major species, *Architeuthis dux*, termed the Atlantic giant squid when differentiated (as it is by some scientists) from a southern, exceptionally long-tentacled version known as *A. longimanus*.

Despite being among the largest invertebrates alive today, virtually nothing is known about the life history of giant squids, but certainly one of their most tantalizing mysteries is their maximum size. According to the record books, the largest giant squid ever recorded was a 55-ft-long specimen beached in Thimble Tickle Bay, Newfoundland, on November 2, 1878. Its body length, measuring from beak tip to tail tip, was 20 feet, and the longest of its appendages was 35 feet. Its largest suckers were four inches across. Zoologists estimate that this molluscan goliath may have weighed several tons. In 1887, an even longer, but far less substantial specimen, classified as *A. longimanus*, was washed ashore on Big Beach, Lyall Bay, in New Zealand's Cook Strait. Although its body length was a mere 7.75 feet, and its weight a modest 300 pounds, its two elongate tentacles each measured a stupendous 54 feet (though these may have been stretched).

However, maritime voyagers and marine biologists alike have frequently claimed far bigger specimens. For instance, several unconfirmed reports of "ultra-giant" squids, estimated to measure in excess of 79 feet, have emerged from around the coasts of Labrador, which is known to be a popular locality for giant squid strandings. Even so, is it possible that such monsters do exist?

Quite apart from anecdotal evidence, tantalizing support has been offered in the form of various sperm whale specimens bearing scars made by squid suckers larger than any reported from known squid species. Of course, it is by no means impossible that there could be relatively modest-sized squids with disproportionately large suckers—but neither should it be impossible for squids, whose hefty body weight is buoyed by their encompassing marine medium, to attain sizes larger than those currently verified by physical remains.

Also needing consideration are individual arms, tentacles, or sections of such structures obtained from the stomachs of whales. As noted by cryptozoologist Bernard Heuvelmans regarding Atlantic specimens in *Dans le Sillage des Monstres Marins: Le Kraken et le Poulpe Colossal* (1958), the dimensions of some of these amputated appendages, when used as a base from which to extrapolate the total length of the entire squids from which they originated, encourage speculation that there may well be squids exceeding 100 feet, and possibly up to:

> ...240 feet long over all. And indeed as female squids have relatively even longer tentacles, some might be more than 300 feet. Even if we knock off 50 per cent to allow for natural exaggeration [and also for the typical elasticity exhibited by squid tentacles and arms] this is a great deal more than the official record of 57 feet.

Perhaps the most amazing modern-day eyewitness account on record of an apparent ultra-giant squid was recalled in 1971 at the annual meeting of the British Association for the Advancement of Science (BA—now the British Science Association) by Ann Bidder, a retired zoology curator from Cambridge University, in support of her belief that giant squids up to at least 200 feet long may exist. The eyewitness was J.D. Starkey, a Royal Navy officer, who had been aboard an Admiralty trawler anchored off one of the Maldives on that remarkable night during World War II when, after hanging a lamp over the side in order to observe the surface marine life, he encountered a very unexpected visitor from the oceanic depths, lying alongside the trawler. In the words of Bidder:

> The light was suddenly obscured and he realized an enormous eye was looking at him. He walked along to the forecastle and saw the ends of tentacles. He went to the stern and there were the other ends of the tentacles [i.e. their bases, attached to the squid's body]. The ship was 175 ft long. A squid with tentacles as long as that would have a body at least 25 ft long.

If Starkey's report was accurate, it means that the giant squid that he encountered was almost four times as long as the Thimble Tickle Bay specimen. And as it was very much alive when viewed, there can be no suggestion that Starkey's specimen had been stretched.

On September 24, 1877, a monstrous multi-limbed sea beast was washed up, still alive, onto Catalina beach in Newfoundland's Trinity Bay. Each of its two great eyes was eight inches across, its body was 9.5 feet in length, and a thrashing mass of tentacles, ten in total and measuring up to 11 feet long, sprouted from its head in dramatic duplication of the legendary hydra. Inevitably, it was dubbed a devil-fish by its eyewitnesses, just like many similar sea monsters washed ashore here and elsewhere during that same period, but today we can readily recognize it as a giant squid.

According to mollusc expert David Heppell, now retired as an invertebrate curator at the Royal Museum of Scotland, however, this huge species may have a much more direct link with the diabolical than merely its potentially terrifying appearance. He has noted that in a beach-stranded giant squid, its tail fin can detach easily from the body and thus present an arrow-shaped or forked outline. A dying squid also releases a pungent, evil-smelling odor, derived from the internal ammonium salts that provide it with buoyancy in life. Moreover, some squids' bodies bear wing-like structures, and many of the larger squids are red or black in color.

Consequently, in a paper presented at Fabulous Beasts: Fact & Folklore, a joint conference of the International Society of Cryptozoology and Britain's Folklore Society that was held at the University of Surrey in Guildford, England, from July 19 to 22, 1990, Heppell speculated that when attempting to reconstruct the likely appearance in life of dead, decomposing giant squids occasionally washed ashore, non-scientific superstitious laymen long ago may well have dreamed up a wholly inaccurate, dramatically different image—converting these colloquially termed devil-fishes into devils. Or is it just a coincidence that the latter demonic entities share the squids' forked tail, wings, red or black skin, and pungent, sulphurous smell?

BLUE PIGS AND URBAN HYAENA-DOGS?

Since joining Facebook in 2008, I have made many new cryptozoological contacts, one of whom is Glenn Cunningham from New York City. During our various FB communications, Glenn has informed me of two separate and very different sightings, 23 years apart, of strange creatures that he has never since forgotten but has never explained to his own satisfaction either. Moreover, these thought-provoking sightings have not been documented in book form until now, so I am most grateful to Glenn for kindly permitting me to document them here.

Sighting #1 occurred 19 years ago, when Glenn was 14 years old, while he was holidaying in Nicaragua with his mother, who hails from this Central American country. Here is the description that he gave to me on March 11, 2011, of what he saw there:

> My mother and I stayed in Esteli, Nicaragua, from mid July 1986 to mid August 1986 visiting family. On the day we were about to set foot on the plane back, in the area of the airport [in Nicaragua's capital city, Managua], there was a large cage with a giant blue pig or hog in it, that had people surrounding it, staring at it in amazement, and of course chattering away...I studied it for 10 minutes...Now, I'm not sure about exactly what I saw, as in, was it a dyed blue pig/hog? But why would they dye a pig, I thought? Anyway, within minutes later we were at the airport, I think by foot too, as it might've been blocks from the airport, and I sat on the plane wondering if blue pigs were rare, or nonexistent, or was that a dyed pig.

There is no known species of wild pig that is even remotely blue in color; and although there are many different breeds of domestic pig, no wholly blue example is on record. Perhaps the closest was a distinctive breed referred to variously as the Bilsdale Blue or the Yorkshire Blue and White. Native to England's North Riding region, it bore large spots of blue pigmentation upon its otherwise white body, and I suppose that it is possible that if in an occasional individual these spots had coalesced to a greater than typical extent, the result would have been a predominantly blue-hued pig.

Even so, the Bilsdale Blue was only a small pig, and certainly not commensurate, therefore, with the "giant" specimen that Glenn witnessed. In any case, such speculation is entirely academic, because sometime prior to the early 1970s this breed had become extinct.

Rather more plausible is that Glenn's perplexing mega-porker had been deliberately dyed blue, but this would surely have endangered the animal's life if the dye had remained on its skin for too long. And for what possible reason would someone choose to dye a pig blue anyway?

In short, the decidedly weird case of the giant blue pig of Nicaragua remains a total enigma, and I know of nothing similar on record. If, conversely, anyone reading this present chapter has at some time observed or has information regarding a blue mystery beast of the porcine persuasion, I'd greatly welcome details.

Glenn's second crypto-sighting occurred much more recently, in 2009, and within sight of his own home in an apartment block almost directly opposite New York City's famous Bronx Zoo. The following description combines a series of accounts that Glenn gave to me on March 11, 2011, in response to various questions of mine regarding the mystery beasts that he witnessed:

> [A] friend and I were parked in front of my building at 2 am talking. We must've been outside the car, cause we wouldn't be able to see backwards, where the Bronx Zoo faces, a couple of blocks from my building. I've seen a pack of raccoons hissing at me in this ghetto concrete buildings area, bats, rabbits (just 3 weeks ago too, but not since 1985 also) and packs of wild dogs in the early to mid 80s. I brought the dogs partly up for a reason obviously. For one, it's been 25 years since I've seen a pack of wild dogs around here. And they were bouncy and friendly actually with their tails high and wagging, at worse, filthy, and ran along on the highway, mainly. On February 2009 my non believing friend and I witnessed a pack of chupacabra-looking dogs (you know, the ones captured on film and found dead since 2006), I think 4, run out of the closed off gated Zoo area. Except they were bigger than the dog-like chupacabras, and kind of hung their shoulders low, and tails were low, you know, like cats do when they're trying to sneak by. The best way to describe them would be, they looked like skinny, high shoulder (but still crouched), bony, ugly hyena/chupacabra dogs that crept across the street into a parking lot where the Transit Authority employees park their cars.
>
> Why this sighting was instantly unique was before I can say anything, my friend who doesn't believe in bigfoot, said aloud: "What the **** is that!" Why would he or I be amazed at regular dogs crossing the street? It had to be something creepy for us to get that impression. Needless to say, after some talking about the encounter I went upstairs and didn't dare look for them. What's even more strange is, I didn't think about it much the next day, and didn't tell anybody or my girlfriend until a month later. Why did I have that kind of nonchalant attitude towards such a weird sighting? Was it a form of shock? I didn't feel shocked the next day, or even that night. I was NUMB though. And yes, a part of me felt like nobody was going to believe me anyway.

The dogs were 3 to 4 feet long. They were big. NOT like the chupacabras exactly. That's why I'm perplexed. More hair possibly too. And the point [where] they were crossing the street it was then one block away from my view. The Zoo's entrance is roughly like 6 blocks away, with the grounds and wooded area a block and a half or 2 away from the front of my building. The back of my building facing that parking area and train L is only a block away at most from the closed off Zoo area. We were closer to the back of my building, for, luckily, better view. A pack of sickly dogs? Hybrid dogs like we discussed?

Seemed to be kind of hairy around the shoulders, but not too much hair elsewhere, Malnourished, UGLY looking, beige, light brownish. Too far [away] for skin color. Skinny all around. Heads too. Maybe like wolf-shaped heads on a shaggy hyena/cabra.

I know, a MESS. Sorry for the bad, all over the place description. I don't know why I use Hyena, as it's probably the wrong dog. Coyotes? Are they skinny and shaggy?

Extremely malnourished Werewolves on Crack. There you have it! That's the best possible description. I think most likely some feral dogs I'm not use to seeing. That's the reality of it. But why have I never seen anything like it before? The strays from the 80s did not look like these.

After sending me these details, Glenn scoured the internet for any animal image that may look something like the creatures that he saw, and eventually he found a photograph of something that did. The creature depicted by the photo was a striped hyaena *Hyaena hyaena*, native to Africa and parts of Asia (this photo can be viewed on my *ShukerNature* blog here: http://www.karlshuker. blogspot.co.uk/2011/03/ blue-pigs-and-hyaena-dogs. html), and Glenn added the following comments in relation to it:

Striped hyaena, a sketch from 1902 (public domain)

The hyena description wasn't as crazy as I thought, this is slightly what I saw, but imagine it crouching, malnourished, and less hair across the spine. No hair on spine like that, but towards shoulder, yes. Hyenas escaped from [Bronx] Zoo? How come I never read about it? Do they return by dawn? Lol

No stripes or spots. Other hyena pics are nothing like what I saw. This one struck my eye as the closest. In fact I was a little shocked when I saw this.

Agreeing with Glenn, I think it highly unlikely that what he saw was a pack of escapee hyaenas, notwithstanding his mystery beasts' superficial similarity to the striped hyaena. Far more reasonable is the assumption that they were merely a foursome of odd-looking feral domestic mongrel dogs, probably the product of several generations of out-breeding, which often yields notably large, vigorous specimens—a phenomenon known as hybrid vigor. (Having said that, however, it is undeniably odd that all four should look the same as one another; canine mongrels are often very dissimilar, even within a single litter.)

Speaking of hybrids, another possibility is that they had resulted from crossbreeding between domestic dogs and some wild canid species, such as the coyote *Canis latrans* (which has resulted in so-called coy-dogs in the past) or the grey wolf *C. lupus* (the ancestor of the domestic dog). When faced with strange-looking canine cryptids, some mystery beast investigators have even speculated whether they could have resulted from hybridization between domestic dogs and hyaenas—which on morphological grounds might seem like a potential explanation for Glenn's mystery beasts. In reality, however, such matings, even if they could occur physically, would not yield viable offspring, because despite their ostensibly canine appearance and behavior, hyaenas constitute a discrete taxonomic family of carnivorous mammals only distantly related to true canids. Indeed, zoologists generally deem hyaenas to be more closely related to cats than to dogs.

Glenn's mystery beasts call to mind another mystifying hyaena-lookalike from North America, known as the shunka warak'in. As I documented in my book *Extraordinary Animals Revisited* (2007):

Translating as 'carrying-off dogs', this is the name given by the Ioway and other native Americans living along the U.S.A.-Canada border to a strange dark-furred creature likened morphologically to a cross between a wolf and a hyaena, which sports a lupine head and high shoulders, but also a sloping back and short hindlimbs—bestowing upon it a hyaenid outline. As its name suggests, the shunka warak'in is said to sneak into the tribes' camps at night and seize any unwary dogs, and it cries like a human if killed.

Glenn has not seen his curious quartet of cryptids again, but somewhere out there, lurking perhaps down some dark alleyway or frequenting some abandoned, derelict shack, there may be a small pack of large, unidentified carnivorous beasts that bear more than a passing resemblance to hyaenas. And that, if nothing else, is a rather disturbing thought.

THE NUNG-GUAMA AND NYAMATSANE—MAN-BEASTS FROM ONCE LONG AGO?

Right from early childhood, I have always greatly enjoyed reading myths, folktales, and legends from around the globe. I was attracted at least in part, no doubt, to the fact that they often included a wide assortment of fantastical fauna, and even as a youngster I rapidly acquired a sizeable collection of books retelling the most famous and also many not-so-famous stories from world folklore.

One of my favorites, which my mother, Mary, bought for me during the mid-1960s and which I still own today, was a wonderful volume entitled *Once Long Ago: Folk and Fairy Tales of the World*. First published in 1962, it contained no less than 70 folktales drawn from cultures across the globe, which were skillfully retold by eminent fantasy writer Roger Lancelyn Green (1918-1987) and beautifully illustrated by Czech artist Vojtěch Kubašta.

Many marvelous beasts of legend and lore appeared in them, including a diverse array of dragons, magical birds, mermaids, a formidable griffin, and Australia's mighty bunyip. In addition to those familiar mythical creatures, however, were two that I had never previously heard of—and, over 40 years after first reading that book, I have still never encountered any notable additional details regarding either of them in any other publication.

Flicking through my much-treasured copy of *Once Long Ago* again recently, I was reminded of these twin enigmas. So I am presenting what little I know of them here, in case anyone else can add to my scant knowledge.

The Nung-Guama—A Chinese Skunk Ape?

Featuring in a Chinese folktale entitled "The Nung-Guama," which told of how a series of kind passers-by helped a poor widow to defeat all attempts by this eponymous monster to devour her, it was succinctly described by its intended victim in the story as follows:

> His body was like a bull, and its head was as big as a wine-jar. His feet were big and soft and floppy: I could hear them going 'Flap-flosh' as he walked. And he had dirty hair and horrid hands with big, sharp claws.

The author holding up his much-treasured copy of *Once Long Ago*, open at the nung-guama story (Karl Shuker)

It was also said to enjoy the taste of human flesh above all other food but was an infamously messy eater; it had a howling "squelchy" speaking voice; it was apparently nocturnal (as it told the widow that it would come back that night to devour her); and it first emerged from out of a clump of bamboo directly in front of the terrified woman. The story was illustrated with several pictures by Kubašta, including two of the nung-guama.

These images are surprisingly reminiscent of the infamous Florida skunk ape (a comparison enhanced by an account elsewhere claiming that it gave off a putrid stink), and they certainly suggest some form of man-beast, perhaps even the Chinese yeren, or at least an ape-like primate.

I know of several other retellings of this tale, including an entire 32-page children's reading book, *The Nung-Guama*, written again by Green, and republished in 1991 by BBC Books with delightful illustrations by Aileen Raistrick. These latter pictures portray it as a savage green biped with tusks and horns—which, at least in my fevered imagination, calls to mind an extraordinary hybrid of Shrek and the Gruffalo!

All of the versions of this folktale that I have seen tell much the same story, but apart from sometimes mentioning that it crunches bones, they offer precious little descriptive details regarding the mystifying nung-guama (or nung-gwama, as spelt in one retelling).

So what is this seemingly-obscure creature? A wholly imaginary Chinese equivalent of Western ogres, hobgoblins, demons, or suchlike? Alternatively, could it be a mythicized form of some known animal species—or even a bona fide cryptid? With so little information to hand, it is extremely risky to attempt any kind of formal identification for the nung-guama—unless there are readers out there who have additional information or can shed further light upon this subject…?

The Nyamatsane—Basuto Mega-Baboon Or Mini Man-Beast?
The second mysterious entity from *Once Long Ago* was another eponymous being, featuring in a Basuto folktale entitled "The Nyamatsanes." In this story, the wife of a brave warrior living in an unnamed arid land (but presumably within Lesotho, formerly known as Basutoland) suddenly developed an insatiable craving for the liver of a nyamatsane. Eventually, the warrior agreed to go out and kill one, and bring back its liver for her to eat. When he reached the place where these creatures lived, however, they had all gone off hunting, except for their very old, feeble grandmother, whom the warrior promptly killed and skinned, after which he cut out her liver. Before he could depart, however, the nyamatsanes returned, so he disguised himself as their grandmother, and succeeded in fooling them, even though they were convinced that they could smell human flesh. When they fell asleep that night, he fled, but although they discovered their dead grandmother's body the next morning and chased after him in fury, he eventually eluded them.

Strangely, however, although the nyamatsanes featured extensively throughout this story, Green's version of it contained no morphological details whatsoever concerning them. Even Kubašta's illustration depicted them only in the most basic, minimalist manner—namely, as dark-colored, tailless, superficially monkey-like entities with featureless faces, predominantly quadrupedal in gait yet also capable of standing and even cavorting bipedally, but shorter in stature than the human warrior. The only details of note that can be gleaned from Green's retelling are that these beings are carnivorous (they were out hunting when the warrior arrived in their domain), they are very attuned to the scent of human flesh, and, very oddly, they are extremely partial to eating pebbles (in order to aid digestion, perhaps?).

The nyamatsanes' folktale appears in similar form within various compendia of African legends, sometimes as the opening section of a much longer story featuring the hare as a trickster, but as with the nung-guama, extra details concerning them are conspicuous only by their absence.

Indeed, the only attempt to categorize the nyamatsane in any way at all that I have so far encountered appears in Jan Knappert's book *Myths and Legends of Botswana, Lesotho and Swaziland* (1985). Included as folktale #43 within a section entitled "The Oldest Tales," during this story's retelling Knappert briefly stated:

> Although the Nyamatsane are supposed to speak fluent if funny Sesotho they are rather pictured as a sort of baboon or man-ape.

This verbal description accords with Kubašta's depiction of them in *Once Long Ago*—the only illustration of nyamatsanes that I have seen. So what could they be?

As suggested for the nung-guama in relation to Chinese myths, they are quite probably nothing more than non-existent entities of African legend, akin to the goblins and other quasi-humanoids of our own traditional folklore. Then again, southern Africa is replete with reports and alleged sightings of diminutive man-beasts likened by Bernard Heuvelmans and various other cryptozoologists to surviving australopithecines, as well as claims of fierce, extra-large monkeys that have been compared with certain species of supposedly extinct giant baboon. Should we therefore be adding the nyamatsanes to the Dark Continent's already-lengthy list of crypto-primates?

Personally, I consider it imperative that before even the most desultory consideration of these beings' possible taxonomic identity is attempted, we obtain substantially more descriptive information concerning them, especially with regard to their morphology and behavior.

Consequently, as with the nung-guama, if anyone out there has such data, I'd like to hear from you—so that the nung-guama and the nyamatsane may ultimately become more than just memories from once long ago.

DID I SEE AN UNDISCOVERED SPECIES OF GIANT PRAYING MANTIS IN SOUTH AFRICA?

The longest species of praying mantis currently known to science is the giant stick mantis *Ischnomantis gigas*. Brown in color, enabling it to blend in with the bushes upon which it lives and lies in wait for unwary prey to approach, this mighty African mantid is native to Senegal, southern Mauritania, Burkina Faso, Mali, northern Nigeria, Cameroon, and Sudan. The longest specimen on record is an adult female collected in Kankiya, northern Nigeria, which measured a very impressive 17.2 cm (6.75 inches) long, and is now preserved in London's Natural History Museum.

Africa is also home to the world's largest mantis species, the aptly named mega-mantis *Plistospilota guineensis*, native to Cameroon, Ivory Coast, Guinea, Liberia, and Ghana. Adult females grow up to 4.3 cm (1.7 inches) long, but are bulkier and heavier (weighing up to 0.35 ounces) than those of the giant stick mantis. They also have much larger wings; the wings of females belonging to the giant stick mantis *I. gigas*, conversely, are so small that the females are rendered flightless.

But could there be even bigger species of mantid *still* awaiting formal scientific discovery and description? The reason why I ask this question is that a few years ago I had a first-hand encounter with a mysterious giant mantis, one that neither I nor anyone else has been able to identify, and which has puzzled me ever since I saw it. So I am now documenting it here, in the hope that someone reading this chapter may be able to offer a solution.

In November 2008, my mother and I spent four days at the private Shamwari Game Reserve, situated just outside Port Elizabeth in South Africa's Eastern Cape. On the last day of our stay there, just a few moments before the car arrived to take us and some other guests to the airport at Port Elizabeth, one of the safari guides walked over towards where we were all waiting, and squatting on the outstretched palm of his right hand was what I can only describe as an absolutely enormous praying mantis.

Brown in color and very burly (hence completely unlike western Africa's notably slender *I. gigas*), this extraordinary specimen was so big that it was easily the length of his entire hand, and it was very much alive. Its "praying" front limbs were moving slightly, and its head

Ischnomantis media, a smaller relative of *I. gigas* (public domain)

turned to look at us as we gazed at it in astonishment. As it made no attempt to fly away, however, I am assuming that it was flightless (and thus a female?).

Frustratingly, my camera was packed away in one of my cases, so I couldn't take any photographs of this amazing insect. Nor could I question the guide about it, because at that same moment the car arrived to take us to the airport, so the guide walked off, still carrying the huge mantis on his hand.

Needless to say, I have never forgotten that spectacular creature, and I have sought ever since to uncover its taxonomic identity, but I have been unable to reconcile it with any mantis species recorded from South Africa—or, indeed, from anywhere else for that matter!

So what *was* this mystery mantis of truly monstrous dimensions? Any suggestions would be greatly welcomed.

THE BEAST OF BUDERIM—IS CRYPTOZOOLOGY'S AUSSIE STAR IN STRIPES A MAINLAND THYLACINE?

During the mid-1990s, many fascinating accounts surfaced online concerning the Beast of Buderim, an unidentified creature reported from Buderim (a mountainous Sunshine Coast region) in Queensland, mainland Australia, which bears an uncanny resemblance to an animal that supposedly died out there over 2,000 years ago.

Take, for example, the sighting made in March 1995 by Buderim dentist Lance Mesh, who spied a strange beast on the fringe of an expanse of rainforest while driving near his home. According to his description, it was: "...goldy, brindly in colour, had a doggish shape and a prominent bump on its head above its eyes." Its most striking feature, however, were the black stripes across its back: "I could not take my eyes off them," said Mesh.

On August 8, 1995, a much more dramatic incident took place, featuring a very similar animal but this time in the vicinity of Bundaberg. Roy Swaby was driving along the main road when suddenly a full-grown male grey kangaroo bounded in front of his vehicle, forcing him to brake heavily in order to avoid hitting it. The kangaroo was evidently fleeing in terror from something—and a few moments later, Swaby discovered what it was:

> This incredible sandy-coloured striped animal leapt out from the
> side of the road a full fifteen feet and into the glare of my 100-watt

halogen spots and four headlights. It stopped on the road, turned to look at me and fell back on to its huge hindquarters, its large green-yellow eyes glowing in the light, and then it opened its jaws and snarled at me. I have never seen anything like it. The white teeth were large and the jaws like a crocodile, like a mantrap. It took two steps and then suddenly crouched and sprang again, 15-20 feet, this time into the scrub...The animal was 4-5 feet long and its huge tail was another 2-3 feet. The stripes started halfway down its back. I thought it was like someone had cut a dingo in half and a 'roo in half and joined them together...On the Thursday following [i.e. August 10] I went to Bundaberg to try to check in the library what it was I'd seen and I found a lithograph of a Tasmanian tiger. There is absolutely no doubt that is what I saw.

Thylacine illustration from 1919 (public domain)

Just like "zebra wolf" and "Tasmanian wolf," "Tasmanian tiger" is one of several colloquial names for Australia's most spectacular species of carnivorous marsupial, *Thylacinus cynocephalus*, the thylacine, which makes it all the more tragic that this remarkable creature is "officially" extinct. Closely resembling a golden-brown wolf or large dog, but patterned across the rear portion of its back and tail with black stripes, on mainland Australia the thylacine suffered greatly from competition with the dingo, introduced by humans, and is believed to have died out there over two millennia ago. On Tasmania, however, it survived until as recently as 1936, when the last fully-confirmed specimen died in Hobart Zoo.

Nevertheless, numerous reports describing thylacine-like beasts have come from Tasmania since then, and it does seem possible that a small population survives in some of this island's wilder, less-explored regions. On the mainland, conversely, such survival would seem far less plausible—were it not for such impressive reports as those given here, and various others like them. What makes these reports so convincing is that their descriptions contain telltale thylacine features that readily discount normal dogs as likely identities.

Although, in evolutionary terms, the thylacine is the marsupials' answer to the wolf, its ancestry is totally separate from that of true wolves and dogs. Hence it exhibits several significant differences. Most noticeable of these are its stripes, and also its jaws. Thylacines could open their mouths to a much wider extent than wolves, yielding an incredible 120° gape, which would certainly explain Swaby's comparison of his mystery beast's jaws with those of a crocodile. Equally unexpected was the thylacine's ability to hop on its hind legs like a kangaroo, but that corresponds perfectly with Swaby's description of his beast as half-dingo, half-kangaroo. Another thylacine idiosyncrasy was a bump above its eyes, which matches the account given by Mesh. Also its long tail was very stiff, far less flexible than a wolf's, and indeed several reports of thylacine-like beasts specifically refer to a stiff, rod-like tail.

Time and again, Queensland and other mainland eyewitnesses have selected the thylacine as the species most similar in appearance and behavior to the striped canine mystery beasts that they have seen, but there is an intriguing twist to this tale of would-be resurrection. The aboriginal people have their own native names for all of Australia's known modern-day animals—but in Queensland they do not appear to have any for the thylacine lookalikes. Yet if these really were native mainland thylacines in this region of the continent, surely they would have their own aboriginal names.

Indeed, there is a notable precedent here, which I documented in my book *Dr Shuker's Casebook* (2008):

> Another dog of the Dreamtime is the marrukurii, which, according to aboriginal traditions prevalent in the vicinity of South Australia's Lake Callabonna, resembled a dog in outline, but was brindled with many stripes. They were believed to be dangerous, especially to human children, carrying away any that they could find to their own special camp at night, where they would savagely devour them. When questioned, the native Australians denied that

the marrukurii were either domestic dogs or dingoes. Is it possible, therefore, particularly in view of their brindled appearance, that these Dreamtime beasts were actually based upon memories of the striped Tasmanian wolf or thylacine *Thylacinus cynocephalus*? After all, this famous dog-like marsupial did not die out on the Australian mainland until about 2300 years ago.

There is, however, one further idea to consider regarding the apparent paradox that if thylacines do indeed exist in Queensland, why do aboriginal people here not have a native name for such creatures? Namely, what if these *are* genuine thylacines, but *not* native mainland specimens? When still common in Tasmania, thylacines were imported onto the Australian mainland as exhibits and even as exotic pets. Did some escape and establish populations in the wild in Queensland? If so, this could uniquely explain not only the current spate of claimed thylacine sightings but also the lack of any native Queensland aboriginal name for them—i.e. their arrival here is too recent for them to have acquired any such name.

Such considerations aside, however, is the Beast of Buderim still being reported today? If so, I'd greatly welcome any information that readers can provide.

A NEW ZEALAND MOA IN CAMBODIA?
They say that beauty is in the eye of the beholder, and the same has certainly been true of cryptids on many occasions in the past. The following case may—or may not—constitute a further example of this cryptozoological rule of thumb.

In terms of their current native zoogeography, modern-day ratites all have very precise distributions on the continental level. The ostrich is nowadays entirely confined to Africa (its contingent in Asia Minor was hunted into extinction by the mid-20th century), the rheas to South America, the emu to Australia, the now-extinct moas to New Zealand, the now-extinct elephant birds to Madagascar, and the cassowaries to Australia and New Guinea. However, there are no known modern-day ratites native to mainland Asia (nor are there any to Europe or North America either, for that matter), which makes a certain enigmatic carving present on a famous Indochinese temple of particular interest.

Dating from the 12th century and richly decorated with countless numbers of bas-relief glyphs carved upon its numerous sandstone columns and walls, depicting a wide range of deities and animals,

Angkor Wat is a celebrated temple complex in Cambodia and constitutes the world's largest religious monument. It also lays claim to cryptozoological fame, courtesy of a specific glyph carved on a wall at Ta Prohm, one of the temples in this complex, because the animal portrayed by this glyph bears a remarkable superficial resemblance to one of the classic plate-backed stegosaurian dinosaurs from prehistoric times. Not surprisingly, this anomalous, ostensibly anachronistic carving has attracted considerable discussion and dissension as to what creature it does truly depict, and I have documented it in a number of my own publications.

However, there is also a second glyph at Angkor Wat that, although far less famous than the "stegosaur," is no less intriguing from a cryptozoological viewpoint because one identity scientifically proposed for the notably long-necked bird that it depicts is a New Zealand moa. This glyph can be found in a temple known as the Bayon, with the mystery bird in question being sandwiched between a carving of a rhinoceros to its immediate left

Bayon glyph depicting rhinoceros, mystery long-necked bird, and ox (public domain)

and one of an ox (possibly a gaur) to its immediate right.

As seen in the illustration of this glyph's animal trio, the bird has stout legs, a noticeably plump winged body, and an extremely long slender neck with a small head atop. In the April 1986 issue of the German scientific periodical *Natur und Museum*, G.H. Ralph von Koenigswald and Joachim Steinbacher correctly pointed out that their morphology ruled out any of the local heron species (the same is true of storks, because both storks and herons possess very long, slender, bayonet-like beaks, whereas the carved bird's is shorter, stouter, and has a hooked tip). They also noted that the reason the glyph carver placed it between two such large mammals as a rhinoceros and an ox (and with its head almost as high as theirs despite the fact that its neck was not even upright but was being held at an angle of approximately 45°) was probably done specifically to demonstrate just how big this bird was.

Reflecting upon these factors, the authors suggested that perhaps the bird was a New Zealand moa, and, if so, quite probably the sturdy, relatively short-legged coastal moa *Euryapteryx curtus* (as opposed to the more famous and taller but much slimmer and longer-legged giant *Dinornis* moas). The moas were not believed to have become extinct in their native New Zealand domain until the mid-1400s (seemingly as a result of over-hunting and habitat destruction by the Maoris), i.e. around 250 years *after* the creation of Angkor Wat. Due to the extensive trade links and maritime travel that had occurred in the southeast Asian-Australasian region for many centuries, the authors believed it likely that New Zealand's mighty moas would have been known about in Indochina at the time of Angkor Wat's creation, and that their spectacular appearance might well have inspired a carving of one to be produced amid the many other depictions of striking wildlife and mythological monsters present here.

Moreover, as the authors also noted, traders throughout history have transported preserved and living specimens of unusual, exotic-looking animals far from their native homelands to those of the traders as curiosities for exhibition purposes. Hence it is remotely possible that merchants travelling between Australasia and Indochina brought a preserved or perhaps even a living moa back with them to Cambodia at some point during the quarter-millennium spanning Angkor Wat's completion and the moas' extinction in New Zealand.

And indeed, there are some very pertinent precedents for transporting living ratites from Australasia to Asia because cassowaries are known to have been transported westwards by mariners in bygone centuries from their native Australian and New Guinea homelands to Indonesia and China. Indeed, as the authors also discussed in this same paper, there is even a glyph of a cassowary-like bird at the Tjandi-Panataran, a Hindu temple not far from Wadjak in Java and dating from around the 12th-15th century, which may offer further evidence of such transportations. Additional details regarding this subject are contained in my book *The Beasts That Hide From Man* (2003).

Glyph of cassowary-like bird at Java's Tjandi-Panataran temple (public domain)

Having said that, there might be an altogether much more mundane, prosaic explanation for the long-necked mystery bird of Angkor Wat. Namely, that its appearance may not be due so much to any taxonomic identity as a moa but rather to the fact that there was a space needing to be filled between the rhino and the ox, and a non-specific long-necked bird simply made an ideal space-filler, with any perceived similarities to *Euryapteryx* or any other moa being merely coincidental. In short, the bird's morphology was molded by the specific shape of the space needing to be filled, nothing more.

Indeed, a telling suggestion that this may well be the case is that whereas the wings of all moas were so small that they were quite probably hidden from view beneath their long hair-like plumage, the Angkor Wat bird has a very large, conspicuous wing readily visible. In addition, moa beaks were not hook-tipped. Such notable discrepancies as these would not be expected if the glyph provides as accurate a representation of the bird as it does for the rhinoceros and the ox, both of which are portrayed realistically and are readily recognizable.

19th-century engraving of common European eels *Anguilla anguilla* emerging from riverbed ooze (public domain)

GIANT BLUE EELS OF THE GANGES—WORMING OUT A POSSIBLE EXPLANATION?

The ancient chroniclers of natural history documented as factual a considerable number of extremely strange, mysterious creatures that are exceedingly implausible from a modern-day zoological standpoint.

Few, however, were more so than the giant worm-like eels with vivid blue bodies that were soberly claimed by Ctesias, Solinus, Philostratus, Aelian, Pliny, and several other famous early scholars to dwell in the dank riverbed ooze of the Ganges and other major Indian rivers.

According to Gaius Iulius Solinus (a renowned Latin scholar and compiler who flourished during the 3rd century AD), these amazing creatures were 30 feet long. However, their dimensions grew ever larger with repeated retellings by later writers, until they eventually acquired sufficient stature—up to 300 feet long now—to emerge from their muddy seclusion beneath the dark cloak of evening and prey upon oxen, camels, and even elephants!

Not surprisingly, this monstrous species of giant freshwater eel has never been brought to scientific attention. True, there are several species of very large sea-dwelling eels, including various morays, that are blue in color. However, there are none known to science that are of comparable size and color but which occur in rivers (interestingly, the longest moray of all, the slender giant moray *Strophidon sathete, is* known from the Ganges and is said to grow up to 13 feet long but is red-grey in color, not blue). So unless the Ganges giant blue eel simply originated with sightings from Asia of sizeable blue marine eels whose correct provenance and dimensions were later documented incorrectly or confused by chroniclers in Europe, then in best angling traditions it is no doubt a classic case of "the one that got away"!

Having said that, however, in recent times I made an interesting discovery that may perhaps provide an alternative core of zoological truth from which the yarn of the giant blue, elephant-engulfing, worm-like Ganges eel was subsequently elaborated and exaggerated.

I discovered that Mount Kinabalu on Borneo is home to a sizeable species of earthworm, measuring up to 28 inches long when fully stretched out, which is iridescent blue in color. Called the Kinabalu giant blue earthworm (but not confined to Borneo, as it also exists on several other nearby southeast Asian islands as well as in New Guinea), it is known scientifically as *Pheretima darnleiensis*.

Moreover, it is such a familiar creature in this region of southern Asia that it is not beyond the realms of possibility that travelers journeying from here to India in bygone times mentioned this eye-catching worm there, and in so doing set the seeds for its transplanted mythification when chronicled in Europe.

Less likely but not impossible is that Asia once harbored a species of blue earthworm rivaling in size those famous giant species native respectively to South Africa and Australia. The largest in Australia is the Gippsland giant earthworm *Megascolides australis*, up to 10 feet long (occasionally more), with a blue-grey body, and to which *Pheretima darnleiensis* just so happens to be closely related. Moreover, Australia is also home to an extremely large species of bright Prussian-blue earthworm, *Terriswalkeris terraereginae*, which can grow up to 6.5 feet long, is native to Queensland's far north, and secretes luminescent mucus. Consequently, sizeable blue earthworms existing in Australasia is by no means unprecedented.

If recollections of a giant blue Asian earthworm or even the smaller *Pheretima darnleiensis* by travelers returning home in Europe became ever more embroidered and distorted with the passing of time, the result might well be a non-existent monster that was not so much a worm-like eel as just a worm, albeit one of unusual, memorable coloration and whose dimensions had become outrageously exaggerated down the generations of retellings.

Thus are legends born.

> *After all, I believe that legends and myths are largely made of "truth," and indeed present aspects of it that can only be received in this mode.*
> — J.R.R. Tolkien, *The Letters of J.R.R. Tolkien*

BIBLIOGRAPHY

ANON. (1869). [Article re supposed American flying toad.] *Grantham Journal* (Grantham), 28 August.

ANON. (1896). Italy's tomb spider. *San Francisco Call* (San Francisco), 29 November.

ANON. (1987). Jaws 5: The eel [Haslar Lake giant conger]. *Daily Mail* (London), 18 September.

ANON. (1992-1993). The vegetable lamb of Tartary. *Fortean Times*, No. 66 (December-January) 34.

ANON. (1994). The beithir...folklore, cryptozoology...or what? *Athene*, No. 4 (September): 9.

ANON. (1994). More thoughts on beithirs and such. *Athene*, No. 5 (Winter): 12-13.

ANON. (1997). Slither of doubt over monster [Gas Street giant eel]. *Express and Star* (Wolverhampton), 5 August.

ANON. (1997). Banks are buzzing with tales of the Brumbeast [Gas Street giant eel]. *Express and Star* (Wolverhampton), 6 August.

ANON. (2010). [News articles re Indonesian giga-gecko]. Trinunnews.com 5 and 6 May.

ANON. (2010). 64-kg gecko sold for $20m: Report. *Jakarta Post* (Jakarta), 8 May.

ANON. (2014). Early summer threatens UK with Volat-Araneus (the flying spider). *Digital Plumbing*, http://www.digitalplumbinguk.com/uncategorized/early-summer-threatens-uk-volat-araneus-flying-spider/ 10 March [no longer online].

ANDREWS, Charles W. (1923). An African chalicothere. *Nature*, 112: 696.

ANDREWS, Roy Chapman (1926). *On the Trail of Ancient Man*. G.P. Putnam's Sons (New York).

ANDREWS, Roy Chapman, *et al.* (1932). *The New Conquest of Central Asia: A Narrative of the Explorations of the Central Asiatic Expeditions in Mongolia and China, 1921-1930.* American Museum of Natural History (New York).

ASCANIUS, Pedr (1772). *Icones Rerum Naturalium, ou Figures Enluminées d'Histoire Naturelle du Nord, Vol. 2.* Claude Philibert (Copenhagen).

AUFFENBERG, Walter (1980). *The Behavioural Ecology of the Komodo Monitor.* University of Florida Press (Gainesville).

AYRE, James (2013). Winged spider hoax—Scientists didn't discover real winged spider. *Planetsave*, http://planetsave.com/2013/07/18/winged-spider-hoax-scientists-didnt-discover-real-winged-spider/ 18 July.

BARBOUR, Erwin H. (1892). Notice of new gigantic fossils [*Daimonelix*]. *Science*, 19: 99-100.

BARBOUR, Erwin H. (1897). Nature, structure and phylogeny of *Daimonelix*. *Bulletin of the Geological Society of America*, 8: 305-314.

BARBOUR, Erwin H. (1903). Present knowledge of the distribution of *Daimonelix*. *Papers in the Earth and Atmospheric Sciences*, No. 349.

BARKER, Will (2003). Nessie is old eunuch eel. *Sun* (London), 22 September.

BECKWITH, Martha (1940). *Hawaiian Mythology*. Yale University Press (New Haven).

BENNETT, Courtenay (1934). Monster or eel? [Réunion's giant eels]. *The Field*, 163 (10 February): 289.

BILLE, Matthew A. (1995). *Rumors of Existence*. Hancock House (Blaine).

BLAUSE, Leticia (2015). A verdade sobre o dodô. *YouTube*, https://www.youtube.com/watch?v=fPkB5Hs_f4E uploaded 13 March.

BONDESON, Jan (1999). *The Feejee Mermaid and Other Essays in Natural and Unnatural History*. Cornell University Press (London).

BONDESON, Jan (2008). The vegetable lamb of Tartary. *Fortean Times*, No. 234 (April): 40-44.

BOUSFIELD, E.L. & LeBLOND, P.H. (1995). An account of *Cadborosaurus willsi*, new genus, new species, a large aquatic reptile from the Pacific coast of North America. *Amphipacifica*, 1, Supplement 1 (20 April): 3-25.

BRUTON, Michael N. (1985). The silver coelacanth. *Ichthos*, 8 (March): 17.

CAGEY1TWO3 (2015). Living dodo captured on video? *YouTube*, https://www.youtube.com/watch?v=dXkikM2M0HE uploaded 3 March.

CARWARDINE, Mark (2007). *Natural History Museum Animal Records*. Natural History Museum (London).

COLEMAN, Loren (2006). See the Jersey devil here!

Cryptomundo, http://cryptomundo.com/cryptozoo-news/norms-devil/ 10 April.

COLEMAN, Loren (2009). Dead bigfoot photo? *Cryptomundo*, http://cryptomundo.com/cryptozoo-news/dead-bigfoot/ 16 April.

COLEMAN, Loren (2009). "Dead bigfoot photo," more. *Cryptomundo*, http://cryptomundo.com/cryptozoo-news/bf-decon2/ 22 April.

[COOK], Steve (2010). A lesson in modern stagecraft. *The Dogman Blog*, http://michigan-dogman.com/wordpress/?p=176 24 March.

[COOK], Steve (2010). The last word. *The Dogman Blog*, http://michigan-dogman.com/wordpress/?p=207 26 March.

COOKE, S.B., *et al.* (2011). An extinct monkey from Haiti and the origins of the Greater Antillean primates. *Proceedings of the National Academy of Sciences*, 108 (No. 7): 2699–2704.

CORRALES, Scott (1997). *Chupacabras and Other Mysteries.* Greenleaf Publications (Murfreesboro).

CZAPLEWSKI, Nicholas J. (2012). A *Mylagaulus* (Mammalia, Rodentia) with nasal horns from the Miocene (Clarendonian) of western Oklahoma. *Journal of Vertebrate Paleontology*, 32 (No. 1): 139-150.

DETHCHEEZ (2006). Mummified alien mothman 2. *Dethcheez*, http://dethcheez.deviantart.com/art/Mummified-Alien-MothMan-2-39107393 2 September.

DETHCHEEZ (2007). Mummified chupacabra fetus 2. *Dethcheez*, http://dethcheez.deviantart.com/art/Mummified-Chupacabra-Fetus-2-54082774 26 April.

DISCOVERY (2010). Giant 'sea serpent' caught on camera [giant oarfish]. *YouTube*, https://www.youtube.com/watch?v=lvRqqwBoyx8 uploaded 9 February.

DURET, Claude (1605). *Histoire Admirable des Plantes et Herbes Esmerueillables et Miraculeuses en Nature.* Nicolas Buon (Paris).

ELLIS, Richard (1998). *The Search for the Giant Squid.* Lyons Press (New York).

EMERY, David (2013). Whoa! Did scientists discover a winged spider? *UrbanLegendsAbout*, http://urbanlegends.about.com/od/spiders/ss/Winged-Spider.htm 21 April.

FRICKE, Hans (1989). Quastie im Baskenland? *Tauchen*, 10 (October): 64-67.

FRICKE, Hans & PLANTE, Raphael (2001). Silver coelacanths from Spain are not proofs of a pre-scientific discovery. *Environmental Biology of Fishes*, 61: 461-463.

GELBART, Mark (2013). The extinct corkscrew beavers of the Miocene. *GeorgiaBeforePeople*, https://markgelbart.wordpress.com/2013/02/20/the-extinct-corkscrew-beavers-of-the-miocene/ 20 February.

GLIDDEN, George (1987). Frog-like monsters attack scientists. *Examiner* [place of publication unknown to me], 11 August.

GOSS, Michael (1986). The great sea-serpent mystery [re giant oarfish]. *The Unknown*, No. 9 (March): 48-51.

GREEN, Roger L. (1962). *Once Long Ago: Folk and Fairy Tales of the World*. Golden Pleasure Books (London).

GRIFFIN, Brian (2013). Have scientists discovered a winged spider? *YouTube*, https://www.youtube.com/watch?v=pXHnrXbf0zc uploaded 15 October.

HART, Martin (1982). *Rats*. Allison and Busby (London).

HARTONO, Rudy (2010). Giant gecko sold [for] at least USD $20,000,000. *My Funny*, http://funfunpics.blogspot.co.uk/2010/05/giant-gecko-sold-at-least-usd-200000.html May.

HAWTAIGNE, Captain (1860). A sea-serpent in the Bermudas. *The Zoologist*, 18: 6934.

HENRY, Teuira & ORSMOND, John M. (1928). Ancient Tahiti. *Bulletin of the Bishop Museum*, No. 48: 1-651.

HEPPELL, David (1990). Was Satan a giant squid? Or the pedigree of the basilisk. Paper presented at 'Fabulous Beasts: Fact & Folklore' [joint conference of the International Society of Cryptozoology and the Folklore Society], 20 July.

HEPPELL, David (1990). Unveiled: Satan's murky origins. *Independent* (London), 30 July.

HEUVELMANS, Bernard (1958). *Dans le Sillage des Monstres Marins: Le Kraken et le Poulpe Colossal*. Plon (Paris).

HEUVELMANS, Bernard (1958). *On the Track of Unknown Animals*. Rupert Hart-Davis (London).

HEUVELMANS, Bernard (1965). Les dragons sont toujours parmi nous. *Atlas*, 6 (October): 76-85.

HEUVELMANS, Bernard (1968). *In the Wake of the Sea-Serpents*. Rupert Hart-Davis (London).

HEUVELMANS, Bernard (1978). *Les Derniers Dragons d'Afrique*.

Plon (Paris).

HEUVELMANS, Bernard (1986). Annotated checklist of apparently unknown animals with which Cryptozoology is concerned. *Cryptozoology*, 5: 1-26.

HEUVELMANS, Bernard (2015). *Les Ours Insolites d'Afrique.* Éditions de l'Oeil du Sphinx (Paris).

HOPKINS, Samantha S.B. (2005). The evolution of fossoriality and the adaptive role of horns in the Mylagaulidae (Mammalia: Rodentia). *Proceedings of the Royal Society B*, 272: 1705-1713.

HUGUES, Albert (1937). Les rois de rats en France. *La Nature*, 65 (1 August): 129.

IMPERATO, Ferrante (1599). *Dell'Historia Naturale.* Costantino Vitale [Felice Stigliola] (Naples).

JORGENSON, Kregg P.J. (2001). *Very Crazy, G.I.: Strange But True Stories of the Vietnam War.* Ballantine (New York).

KEELING, Clinton H. (1995). The British Nandi bear? *Animals and Men*, No. 6 (July): 32-33.

KNAPPERT, Jan (1985). *Myths and Legends of Botswana, Lesotho and Swaziland.* E.J. Brill (Leiden).

KOENIGSWALD, G.H. Ralph von & STEINBACHER, Joachim (1986). Fremde Vögel an fernem Ort. *Natur und Museum*, 116 (April): 97-103.

LeBLOND, Paul H. & BOUSFIELD, Edward L. (1995). *Cadborosaurus: Survivor From the Deep.* Horsdal and Schubart (Victoria).

LEAKEY, L.S.B. (1935). Does the chalicothere—contemporary of the okapi—still survive? *Illustrated London News*, 97 (2 November): 730-733, 750.

LEAKEY, L.S.B. (1942). Fossil Suidae from Oldoway. *Journal of the East Africa and Uganda Natural History Society*, 16: 178-190.

LEE, Henry (1887). *The Vegetable Lamb of Tartary: A Curious Fable of the Cotton Plant.* Sampson Low, Marston, Searle, and Rivington (London).

LEY, Willy (1945). The quest for Africa's unknown animals. *Travel*, 85 (October): 16-19, 32.

LEY, Willy (1959). *Exotic Zoology.* Viking Press (New York).

LEY, Willy (1963). Is there a Nandi bear? *Fate*, 16 (July): 42-50.

LEY, Willy (1963). King of the rats. *Galaxy*, 22 (October): 85-91.

LOVERIDGE, Arthur (1945). African native attacked by a giant frog. *Copeia*, (31 December): 232.

LOXTON, Daniel (2009). The shocking secret of Thetis Lake. *Junior Skeptic*, No. 35. *In*: *Skeptic*, 15 (No. 2): 74-81.

MacPHEE, R.D.E. (1996). The Greater Antillean monkeys. *Revista de Ciència (Institut d'Estudis Baleàrics)*, 18: 13-32.

MacPHEE, R.D.E. & FLEAGLE, J.G. (1991). Postcranial remains of *Xenothrix mcgregori* (Primates, Xenotrichidae) and other Late Quaternary mammals from Long Mile Cave, Jamaica. *Bulletin of the American Museum of Natural History*, 206: 287-321.

MACKAL, Roy P. (1987). *A Living Dinosaur? In Search of Mokele-Mbembe.* E.J. Brill (Leiden).

MARDIS, Scott (2013). What was the Naden Harbour carcass aka Cadborosaurus willsi? *Bizarre Zoology*, http://bizarrezoology.blogspot.co.uk/2013/06/what-was-naden-harbor-carcass-part-1-of.html 17 June.

MARTIN, Larry D. (1994). The devil's corkscrew. *Natural History*, 103 (April): 59-60.

MARTIN, Larry D. & BENNETT, D.D. (1977). The burrows of the Miocene beaver *Palaeocastor*, Western Nebraska, U.S.A. *Palaeogeography, Palaeoclimatology, Palaeoecology*, 22: 173-193.

MATTHEW, William D. (1902). A horned rodent from the Colorado Miocene; with a revision of the Mylagauli, beavers, and hares of the American Tertiary. *Bulletin of the American Museum of Natural History*, 16: 291-310.

McCLOY, James F & MILLER, Ray (1976). *The Jersey Devil.* Middle Atlantic Press (Moorestown).

McRAE, Toni (1995). Beast of Buderim. *Sunday Mail* (Brisbane), 18 June.

McRAE, Toni (1995). Odd beast seen again. *Sunday Mail* (Brisbane), 25 June.

McRAE, Toni (1995). Beast of Buderim on the prowl. *Sunday Mail* (Brisbane), 2 July.

MICHELL, John & RICKARD, Robert J.M. (1982). *Living Wonders: Mysteries and Curiosities of the Animal World.* Thames and Hudson (London).

MILJUTIN, Andrei (2007). Rat kings in Estonia. *Proceedings of the Estonian Academy of Sciences: Biology and Ecology*, 56 (No. 1): 77–81.

MILTON, Giles (1996). *The Riddle and the Knight: In Search of*

Sir John Mandeville. Allison and Busby (London).

MOHR, Erna (1929). Ein "Rattenkönig" von Waldmäusen. *Zeitschrift für Säugetierkunde*, 4: 252.

MOORE, Robin (2014). *In Search of Lost Frogs: The Quest to Find the World's Rarest Amphibians.* Bloomsbury Natural History (London).

NAISH, Darren (1997). Another Caddy carcass? *The Cryptozoology Review*, 2 (Summer): 26-29.

NAISH, Darren (2015). People are modifying monitors to make gargantuan geckos. *Tetrapod Zoology*, http://blogs.scientificamerican.com/tetrapod-zoology/people-are-modifying-monitors-to-make-gargantuan-geckos/ 16 April.

NG, Chris (2009). The legend of Michigan's dogman Gable Film full (HD). *YouTube*, https://www.youtube.com/watch?v=_jQYbgEZrTA uploaded 15 July.

PETERSON, O.A. (1905). Description of new rodents and discussion of the origin of *Daemonelix. Memoirs of the Carnegie Museum*, 2: 139-200.

PICKFORD, Martin (1975). Another African chalicothere. *Nature*, 253 (10 January): 85.

PITMAN, Charles R.S. (1931). *A Game Warden Among His Charges.* James Nisbet (London).

POCOCK, Reginald I. (1930). The story of the Nandi bear. *Natural History Magazine*, 2: 162-169.

POISSON, R. & PESSON, P. (1937). Présentation d'un "Roi de chats" observé à Rennes. *Bulletin de la Societé Scientifique de Bretagne*, 14 (May): 186-188.

PROSEK, James (2010). Maori eels. *Orion Magazine*, 29 (No. 4; July-August) [https://orionmagazine.org/article/survivors/].

RADFORD, Benjamin (2011). *Tracking the Chupacabra: The Vampire Beast in Fact, Fiction, and Folklore.* University of New Mexico Press (Albuquerque).

RADFORD, Benjamin (2013). Jersey devil: Impossible animal of story and legend. *Live Science*, http://www.livescience.com/28167-jersey-devil.html 25 March.

RATELIFF, John D. (Ed.) (2007). *The History of The Hobbit, Parts One and Two.* HarperCollins (London).

REDFERN, Nick (2008). The great eel of Birmingham. *There's Something in the Woods...*, http://monsterusa.blogspot.co.uk/2008/04/great-eel-of-birmingham.html 18 April.

REH, A. (1937). Les "rois de rats". *La Nature*, 65 (15 June): 537-540.

RICKARD, Robert J.M. (1983). Rat kings. *Fortean Times*, No. 40 (Summer): 10-11.

ROBERTS, Tyson R. (2012). *Systematics, Biology, and Distribution of the Species of the Oceanic Oarfish Genus Regalecus (Teleostei, Lampridiformes, Regalecidae).* Muséum National d'Histoire Naturelle (Paris).

ROTHSCHILD, L.W. (1905). Exhibition of two tusks from Abyssinia. *Proceedings of the Zoological Society of London*, 2 [for 1905] (14 November): 297.

ROTHSCHILD, Maurice de & NEUVILLE, Henri (1907). Sur une dente d'origine énigmatique. *Archives de Zoologie Expérimentale et Générale, Série 4, 7* (15 October): 270-333.

SALUSBURY, Matt (2013). *Pygmy Elephants: On the Track of the World's Largest Dwarfs.* CFZ Press (Bideford).

SALUSBURY, Matt (2014). Can anyone help identify this "tooth (tusk?) of enigmatic origin"? *Matt Salusbury*, http://mattsalusbury. blogspot.it/2014/06/can-anyone-help-identify-this-tooth.html 6 June.

SAMBUCUS, Johannes (1576). *Emblemata, Cum Aliquot Nummis Antiqui Operis* (4th edition). Plantin Press (Antwerp).

SÁNCHEZ-OCEJO, Virgilio (1997). *Miami "Chupacabras".* Pharaoh Production (Miami).

SELLECK, Murray K., *et al.* (1995). A morbillivirus that caused fatal disease in horses and humans. *Science*, 268: 94-97.

SHUKER, Karl P.N. (1995). *In Search of Prehistoric Survivors: Do Giant 'Extinct' Creatures Still Exist?* Blandford (London).

SHUKER, Karl P.N. (1996). *The Unexplained: An Illustrated Guide to the World's Natural and Paranormal Mysteries.* Carlton (London).

SHUKER, Karl P.N. (1999). *Mysteries of Planet Earth: An Encyclopedia of the Inexplicable.* Carlton (London).

SHUKER, Karl P.N. (2003). *The Beasts That Hide From Man: Seeking the World's Last Undiscovered Animals.* Paraview (New York).

SHUKER, Karl P.N. (2007). *Extraordinary Animals Revisited: From Singing Dogs To Serpent Kings.* CFZ Press (Bideford).

SHUKER, Karl P.N. (2008). *Dr Shuker's Casebook: In Pursuit of Marvels and Mysteries.* CFZ Press (Bideford).

SHUKER, Karl P.N. (2012). *The Encyclopaedia of New and Rediscovered Animals: From The Lost Ark To The New Zoo—And Beyond.* Coachwhip Publications (Landisville).

SHUKER, Karl P.N. (2013). *Mirabilis: A Carnival of Cryptozoology and Unnatural History.* Anomalist Books (San Antonio).

SHUKER, Karl P.N. (2013). *Dragons in Zoology, Cryptozoology, and Culture.* Coachwhip Publications (Greenville).

SHUKER, Karl P.N. (2014). *The Menagerie of Marvels: A Third Compendium of Extraordinary Animals.* CFZ Press (Bideford).

SIEVEKING, Paul (1992). Tangled tales [squirrel kings]. *Fortean Times*, No. 63 (June-July): 13.

SIEVEKING, Paul (1997). Tangled tales of the squirrels in knots. *Daily Telegraph* (London), 3 August.

SIEVEKING, Paul (1997). Twisted tails of squirrels. *Fortean Times*, No. 104 (November): 11.

SKERRY, Brian (1997). High-seas drifter [giant oarfish]. *BBC Wildlife*, 15 (June): 64-65.

SLOANE, Hans (1698). A further account of the contents of the China cabinet mentioned last Transaction, p. 390. *Philosophical Transactions of the Royal Society*, 20 (1 January): 461-462.

SLOANE, Hans (1707, 1725). *A Voyage to the Islands Madera, Barbados, Nieves, S. Christophers and Jamaica, Vols 1, 2.* B.M. (London).

SLOANE, Hans (1725). [Vegetable lamb of Tartary]. *Philosophical Transactions of the Royal Society*, 33: 353.

SMITH, Andrew (1877). [Letter re giant oarfish as sea serpent]. *The Times* (London), 15 June.

SNOPES (2014). Scientist discovers winged spider. *Snopes*, http://snopes.com/photos/bugs/wingedspider.asp March.

SPEIGHT, W.L. (1940). Mystery monsters in Africa. *Empire Review*, 71: 223-228.

STARKEY, J.D. (1963). 'I saw a sea monster'. *Animals*, 2 (No. 23; 23 September): 629, 644.

STEFFEN, Chad (1996). Water monitors. *Reptiles*, (June): 68-69.

STOW, George W. & BLEEK, Dorothea F. (1930). *Rock-Paintings in South Africa From Parts of the Eastern Province and Orange Free State.* Methuen (London).

SWANCER, Brent (2014). Giant mystery fish of Japan. *Mysterious Universe*, http://mysteriousuniverse.org/2014/04/giant-mystery-fish-of-japan/ 30 April.

SYLVA, Donald P. de (1966). Mystery of the silver coelacanth. *Sea Frontiers*, 12 (May-June): 172-175.

TAYLOR, Paul D. & SENDINO, Consuelo (2009). *Dinocochlea*: the mysterious spiral of Hastings. *Deposits*, No. 20 (Autumn): 40–41.

TAYLOR, Paul D. & SENDINO, Consuelo (2011). A new hypothesis for the origin of the supposed giant snail *Dinocochlea* from the Wealden of Sussex, England. *Proceedings of the Geologists' Association*, 122 (June): 494-500.

THESIGER, Wilfred (1964). *The Marsh Arabs*. Longmans (London).

TROUESSART, Emile (1911). Existe-t-il dans les marais du Lac Tchad un grand mammifère encore inconnu des naturalistes? *La Nature*, 76 (21 January): 120-121.

TYREE, James S. (2010). Ancient horned rodent creates museum excitement 50 years later. *The Oklahoman* (Oklahoma City), 28 December.

UN-CRUISE ADVENTURES (2014). Oarfish. *YouTube*, https://www.youtube.com/watch?v=IoDqG0syBFE uploaded 24 March.

WAHID, Abdul (2010). Primacy effect. *Suparwi's Blog*, http://ciucc.wordpress.com/2010/05/15/primacy-effect 15 May.

WEIRDTHEATRE (2009). Gable Film Part 2. *YouTube*, https://www.youtube.com/watch?v=CShNKGRY5tw uploaded 23 July.

WESTERVELT, W.D. (1916). *Hawaiian Legends of Ghosts and Ghost-Gods*. Ellis Press (Boston).

WHITTALL, Austin (2010). Horse eating frog—Chile. *Patagonian Monsters*, http://patagoniamonsters.blogspot.co.uk/2010/11/horse-eating-frog-chile.html 9 November.

WIANECKI, Shannon (2012). The sacred spine [re Hawaiian mo'o lizards]. *Maui Magazine*, http://www.mauimagazine.net/Maui-Magazine/September-October-2012/The-Sacred-Spine/ (September-October), 5 pp.

WILLIAM OF SWEDEN, Prince (1923). *Among Pygmies and Gorillas With the Swedish Zoological Expedition to Central Africa 1921*. Gyldendal (London).

WILLIAMS, E.E. & KOOPMAN, K.F. (1952). West Indian fossil monkeys. *American Museum Novitates*, No. 1546, 16 pp.

WOODWARD, B.B. (1922). On *Dinocochlea ingens*, n. gen. et sp., a gigantic gastropod from the Wealden Beds near Hastings.

Geological Magazine, 59: 242–247.

    WOOLDRIDGE, Ian (1987). Quest for the conger [Haslar Lake giant conger]. *Daily Mail* (London), 17 September.

    WOOLHEATER, Craig (2006). Another dead bigfoot photo? *Cryptomundo*, http://cryptomundo.com/bigfoot-report/dead-bigfoot-photo2/ 21 November.

ACKNOWLEDGEMENTS

As ever, a great many people were of assistance to me when preparing this book, but in particular the following:

David Alderton, the late Mark K. Bayless, Matt Bille, Janet Bord, Dr Ed Bousfield, Markus Bühler, Loren Coleman, Glenn Cunningham, Bob Deis, Jonathan Downes/CFZ, Thomas Finley, *Fortean Times*, Emily Fuggle/Garden Museum, Bill Gibbons, the late J. Richard Greenwell/International Society of Cryptozoology, David Heppell, 'Justathought', the late Clinton Keeling, Connor Lachmanec, Prof. Paul LeBlond, Håkan Lindh, Cameron A. McCormick, the late Prof. Roy P. Mackal, the late Ivan Mackerle, Scott Mardis, Tim Morris, Richard Muirhead, Bill Munns, Dr Darren Naish, Tony Nichol, Hodari Nundu, Brian D. Parsons, Martin Phillipps, *Practical Reptile Keeping*, Richard Pullen, Benjamin Radford, Michel Raynal, William M. Rebsamen, Chris R. Richards, Matt Salusbury, the late Mary D. Shuker, Bob Skinner, *Small Furry Pets*, Richard Svensson, Gerard Van Leusden, Paolo Viscardi, Richard S. White, the late Gerald L. Wood, Craig Woolheater.

I also wish to offer an especial vote of thanks to my editor and publisher Patrick Huyghe at Anomalist Books for his continuing interest and enthusiasm with regard to my researches and writings; to Ken Gerhard for his very kind and inspirational foreword; and to Michael J. Smith for the wonderful artwork that not only constitutes the spectacular front cover illustration for this book but also was the original inspiration for the entire book's conception.

ABOUT THE AUTHOR

Born and still living in the West Midlands, England, Karl P.N. Shuker graduated from the University of Leeds with a Bachelor of Science (Honours) degree in pure zoology, and from the University of Birmingham with a Doctor of Philosophy degree in zoology and comparative physiology. He now works full-time as a freelance zoological consultant to the media, and as a prolific published writer.

Shuker is currently the author of 22 books and hundreds of articles, principally on animal-related subjects, with an especial interest in cryptozoology and animal mythology, on which he is an internationally recognized authority, but also including a poetry volume. In addition, he has acted as consultant for several major multi-contributor volumes as well as for the world-renowned *Guinness Book of Records/Guinness World Records* (he is currently its Senior Consultant for its Life Sciences section); and he has compiled questions for the BBC's long-running cerebral quiz *Mastermind*. He is also the editor of the *Journal of Cryptozoology*, the world's only existing peer-reviewed scientific journal devoted to mystery animals.

Shuker has travelled the world in the course of his researches and writings, and has appeared regularly on television and radio. Aside from work, his diverse range of interests include motorbikes, the life and career of James Dean, collecting masquerade and carnival masks, quizzes, philately, poetry, travel, world mythology, and the history of animation.

He is a Scientific Fellow of the prestigious Zoological Society of London, a Fellow of the Royal Entomological Society, a Member of the International Society of Cryptozoology and other wildlife-related organizations, he is Cryptozoology Consultant to the Centre for Fortean Zoology, and is also a Member of the Society of Authors.

Shuker's personal website can be accessed at http://www.karlshuker. com and his mystery animals blog, *ShukerNature*, can be accessed at http://www.karlshuker.blogspot.com

His poetry blog can be accessed at http://starsteeds.blogspot.com and his *Eclectarium* blog can be accessed at http://eclectariumshuker. blogspot.com

There is also an entry for Shuker in the online encyclopedia Wikipedia at http://en.wikipedia.org/wiki/Karl_Shuker and a Like (fan) page on Facebook.

AUTHOR BIBLIOGRAPHY

Mystery Cats of the World: From Blue Tigers To Exmoor Beasts (Robert Hale: London, 1989)

Extraordinary Animals Worldwide (Robert Hale: London, 1991)

The Lost Ark: New and Rediscovered Animals of the 20th Century (HarperCollins: London, 1993)

Dragons: A Natural History (Aurum: London/Simon & Schuster: New York, 1995; republished Taschen: Cologne, 2006)

In Search of Prehistoric Survivors: Do Giant 'Extinct' Creatures Still Exist? (Blandford: London, 1995)

The Unexplained: An Illustrated Guide to the World's Natural and Paranormal Mysteries (Carlton: London/JG Press: North Dighton, 1996; republished Carlton: London, 2002)

From Flying Toads To Snakes With Wings: From the Pages of FATE Magazine (Llewellyn: St Paul, 1997; republished Bounty: London, 2005)

Mysteries of Planet Earth: An Encyclopedia of the Inexplicable (Carlton: London, 1999)

The Hidden Powers of Animals: Uncovering the Secrets of Nature (Reader's Digest: Pleasantville/Marshall Editions: London, 2001)

The New Zoo: New and Rediscovered Animals of the Twentieth Century [fully-updated, greatly-expanded, new edition of *The Lost Ark*] (House of Stratus Ltd: Thirsk, UK/House of Stratus Inc: Poughkeepsie, USA, 2002)

The Beasts That Hide From Man: Seeking the World's Last Undiscovered Animals (Paraview: New York, 2003)

Extraordinary Animals Revisited: From Singing Dogs To Serpent Kings (CFZ Press: Bideford, 2007)

Dr Shuker's Casebook: In Pursuit of Marvels and Mysteries (CFZ Press: Bideford, 2008)

Dinosaurs and Other Prehistoric Animals on Stamps: A Worldwide Catalogue (CFZ Press: Bideford, 2008)

Star Steeds and Other Dreams: The Collected Poems (CFZ Press: Bideford, 2009)

Karl Shuker's Alien Zoo: From the Pages of Fortean Times (CFZ Press: Bideford, 2010)

The Encyclopaedia of New and Rediscovered Animals: From The Lost Ark to The New Zoo—and Beyond (Coachwhip Publications: Landisville, 2012)

Cats of Magic, Mythology, and Mystery: A Feline Phantasmagoria (CFZ Press: Bideford, 2012)

Mirabilis: A Carnival of Cryptozoology and Unnatural History (Anomalist Books: San Antonio, 2013)

Dragons in Zoology, Cryptozoology, and Culture (Coachwhip Publications: Greenville, 2013)

The Menagerie of Marvels: A Third Compendium of Extraordinary Animals (CFZ Press: Bideford, 2014)

A Manifestation of Monsters: Examining the (Un)Usual Suspects (Anomalist Books: San Antonio, 2015)

Consultant and also Contributor

Man and Beast (Reader's Digest: Pleasantville, New York, 1993)

Secrets of the Natural World (Reader's Digest: Pleasantville, New York, 1993)

Almanac of the Uncanny (Reader's Digest: Surry Hills, Australia, 1995)

The Guinness Book of Records/Guinness World Records 1998-present day (Guinness: London, 1997-present day)

Consultant

Monsters (Lorenz: London, 2001)

Contributor

Of Monsters and Miracles CD-ROM (Croydon Museum/Interactive Designs: Oxton, 1995)

Fortean Times Weird Year 1996 (John Brown Publishing: London, 1996)

Mysteries of the Deep (Llewellyn: St Paul, 1998)

Guinness Amazing Future (Guinness: London, 1999)

The Earth (Channel 4 Books: London, 2000)

Mysteries and Monsters of the Sea (Gramercy: New York, 2001)

Chambers Dictionary of the Unexplained (Chambers: Edinburgh, 2007)

Chambers Myths and Mysteries (Chambers: Edinburgh, 2008

The Fortean Times Paranormal Handbook (Dennis Publishing: London, 2009)

Plus numerous contributions to the annual *CFZ Yearbook* series of volumes, and also to the annual *Fortean Studies* series of volumes.

Editor

The *Journal of Cryptozoology* (CFZ Press: Bideford, 2012-present day)

INDEX OF ANIMALS AND PLANTS